PUZZLE BOOK

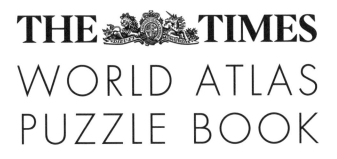

THE TIMES WORLD ATLAS PUZZLE BOOK

THE TIMES WORLD ATLAS PUZZLE BOOK
Published by Times Books
An imprint of HarperCollins Publishers
Westerhill Road
Bishopbriggs
Glasgow G64 2QT
harpercollins.co.uk

First published 2019

© HarperCollins Publishers 2019
Maps © Collins Bartholomew Ltd 2019
Puzzles © Any Puzzle Media Limited

Puzzles by Gareth Moore

The Times® is a registered trademark of Times Newspapers Ltd

A catalogue record for this book is available from the British Library.

ISBN 978-0-00-835178-6

10 9 8 7 6 5 4 3 2

Printed in Slovenia

All mapping in this book is generated from Collins Bartholomew digital databases. Collins Bartholomew, the UK's leading independent geographical information supplier, can provide a digital, custom, and premium mapping service to a variety of markets. For further information:
Tel: +44 (0)208 307 4515
e-mail: collinsbartholomew@harpercollins.co.uk
or visit our website at: www.collinsbartholomew.com

If you would like to comment on any aspect of this book, please contact us at the above address or online.
timesatlas.com
e-mail: timesatlas@harpercollins.co.uk

CONTENTS

INTRODUCTION TO THE TIMES ATLAS

The Times Atlas of the World has been published for over a hundred years and is regarded as the world's most prestigious and authoritative atlas.

The first edition was published in 1895, with maps produced by the German firm Velhagen & Klasing. The Times Publishing Co Ltd then commissioned John Bartholomew & Son, the renowned Edinburgh cartographers, to prepare a revised world atlas which became *The Times Survey Atlas & Gazetteer of the World* (1922). The atlas set new standards for cartography and included an Index-Gazetteer which itself became an authority on the place names of the world.

Bartholomew & Son pushed cartographic design to new levels of excellence and introduced new production techniques that revolutionised mapping. The company was first bought by Reader's Digest in 1980 and then by News International in 1985. Finally, in 1989, it became part of HarperCollins Publishers and publication of Times atlases and maps continued with *The Times World Political Map and World Physical Map* (1989).

The introduction of digital technology to produce atlases in the nineties marked a huge change within the cartographic profession. Today, the long-standing connection with Bartholomew & Son continues with the latest family of atlases, which use the authoritative Collins Bartholomew world and regional geographic databases to create each map plate. These seamless databases allow map projection and scale parameters to be individually selected to give the optimum presentation.

The Times Atlas of the World takes advantage of technological advances and the increasing availability of geographic information – neither of which could have been envisaged by the cartographers who produced the earliest edition – to ensure it retains the integrity, authority, beauty and tradition for which Times atlases are renowned. By doing so, it provides the most comprehensive portrayal of the world available in any reference atlas.

MAP SYMBOLS

PLACE NAMES

In this puzzle book, conventional names in English (where they exist) are used in all questions and answers for sections 1 and 2. For example, Moscow is referred to in the conventional form, rather than Moskva, which is the romanized form from Russian Cyrillic (Москва). In section 3, the mapping uses the local spelling with the English conventional form in brackets, if one exists, and this convention is reflected in the questions and answers to this section. There may be some exceptions for very well-known English-language names, such as Suez Canal, where the question uses the English conventional form, but the map uses the local name as the main form (Qanāt as Suways).

BOUNDARIES

The status of nations and their names and boundaries are shown in this puzzle book as they are at the time of going to press, as far as can be ascertained, including very recent changes. Where international boundaries are disputed, it may be that no portrayal of them will meet with the approval of any of the countries involved. It is not the purpose of the *The Times Atlas of the World* or this book to adjudicate on geographical issues, and reference mapping at atlas scales is not the ideal medium for indicating the claims of separatist and irredentist movements. However, every reasonable attempt is made to show where an active territorial dispute exists, and where there is an important difference between 'de facto' (existing in fact, on the ground) and 'de jure' (according to law) boundaries. This is done by using a different symbol for international boundaries that are disputed, or where the alignment is unconfirmed, from that used for settled international boundaries. Ceasefire lines are also shown by a separate symbol. Where necessary, annotation is provided for clarity. The atlas aims to take a strictly neutral viewpoint of all such cases, based on advice from expert consultants.

MAP SYMBOLS

Cities and towns

Population	National capital	Administrative capital Shown for selected countries only First order	Second order Scales larger than 1:9 000 000	Other city or town
over 10 million	TŌKYŌ ▣	Karachi ▣	Los Angeles ☉	New York ☉
5 million to 10 million	SANTIAGO ▣	Nanjing ▣	Dongguan ☉	Chicago ☉
1 million to 5 million	KĀBUL ▣	Sydney ▣	Tangshan ◉	Puning ◉
500 000 to 1 million	BANGUI ▣	Mykolayiv ▣	Warangal ◎	Raulakela ◎
100 000 to 500 000	WELLINGTON ▢	Rangpur ▢	Naogaon ⊙	Apucarana ⊙
50 000 to 100 000	PORT OF SPAIN ▢	Potenza ▢	Bayreuth ○	Tuxpan ○
10 000 to 50 000	KINGSTOWN ▫	Trujillo ▫	Willimantic ○	Ceres ○
1 000 to 10 000	VADUZ ▫	Ati ▫	Nepalganj ○	Abla ○
under 1 000 Scales 1:3 000 000 and larger	NGERULMUD ▫	Melekeok ▫	Carmel ○	Lopigna ○

Styles of lettering

Cities and towns are explained separately

Physical features and areas

Country	**FRANCE**	*Mt Blanc*
Overseas Territory/Dependency	**Guadeloupe**	*Thames*
Disputed Territory	WESTERN SAHARA	*Okavango Delta*
Administrative name, first order internal division Shown for selected countries only	SCOTLAND	*Gran Canaria*
Administrative name, second order internal division Scales 1:3 000 000 and larger Shown for selected countries only	MANCHE	*LAKE ERIE*
Geographical area	ARTOIS	*PAMPAS*
Historical area	LYDIA	*ANDES*

Boundaries

International boundary

Disputed international boundary or alignment unconfirmed

Undefined international boundary in the sea. All land within this boundary is part of state or territory named

Ceasefire line or other boundary described on the map

Disputed territory boundary

UN buffer zone

Administrative boundary, first order internal division
Scales 1:3 000 000 and larger
Shown for selected countries only

Administrative boundary, first order internal division
Scales smaller than 1:3 000 000
Shown for selected countries only

Administrative boundary, second order internal division
Scales 1:3 000 000 and larger.
Shown for selected countries only

Disputed administrative boundary
Scales 1:3 000 000 and larger
Shown for selected countries only

Transport

Motorway
Scales 1:3 000 000 and larger

Main road

Secondary road

Motorway tunnel

Road tunnel

Track

Main railway

Secondary railway

Railway tunnel

Canal

Minor canal

Main airport, regional airport

Land and sea features

Rock desert

Sand desert / Dunes

Oasis

Lava field

1234 Volcano
Height in metres

Marsh

Ice cap / Glacier

Nunatak

Coral reef

Escarpment

Flood dyke

) (123 Pass
Height in metres

Ice shelf

2000 Ice surface elevation
above sea level in metres

1234 Summit
Height in metres

2835 Spot height
Surface height in metres
for depressions, areas
below sea level and other
points above sea level

5678 Ocean deep
In metres
Ocean plates only

Lakes and rivers

Lake

Impermanent Lake

Salt lake or lagoon

Impermanent salt lake

Dry salt lake or salt pan

123 Lake height
Surface height above sea level, in metres

River

Impermanent river

Wadi or watercourse

Waterfall

Dam

Barrage

Other features

National park

Reserve or special land area

Site of specific interest

Wall

Built up area

9

HOW TO USE THIS BOOK

The puzzles in this book have been carefully selected to showcase the fascinating and exhaustive range of information held in the Collins Bartholomew world and geographic databases. You will need to use your wits, cunning and general knowledge to solve problems on topics as diverse as time zones, country names, capital cities and Antarctica.

We have organized the quizzes into three areas. The World Today will challenge the reader on population, climate and health. Geographical Information will present you with currency conundrums and brain teasers on everything from mountains to capital cities to US states. Finally, the continental maps present a selection of cryptic problems, anagrams and puzzles about specific regions.

Some of the quizzes use maps and data from the *Times Comprehensive Atlas of the World*, *Times Universal Atlas of the World* and the *Times Concise Atlas of the World* to encourage readers to interrogate the information in fun and thought-provoking new ways. For other puzzles, we have specially developed new maps to test your understanding of the world.

No matter your geographical know-how, there is something here for everyone. Some tasks will require more general knowledge than others, but we hope that all the puzzles in this book will entertain, challenge and inspire you to look at our maps with renewed interest and to use them as a starting point to explore more of the fascinating world in which we live.

Good luck!

PUZZLES

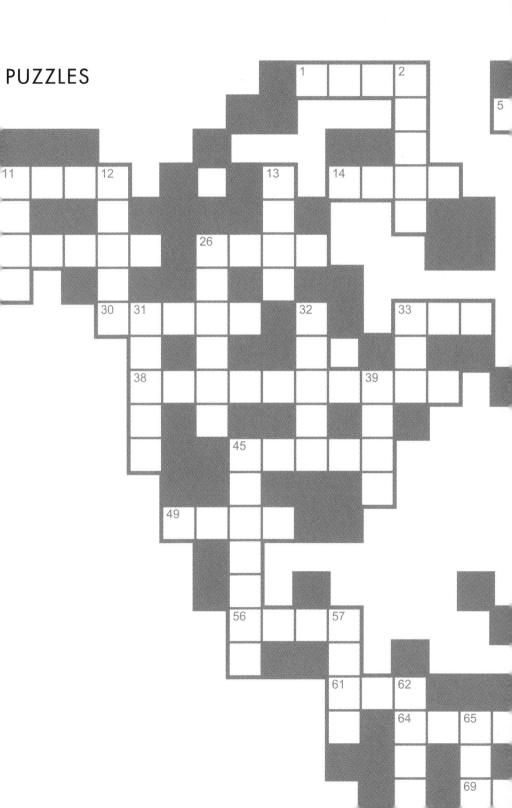

THE WORLD TODAY

1 WEATHER

The colourful map opposite reveals the world's major climatic regions, along with information about tropical storms over a period of two decades.

1 According to the map, in which year(s) did the greatest number of major tropical storms occur?

2 What is the most common climatic region at the equator?

3 What features do all of the tropical storms share?

4 Match the tropical storm type to the region of the world in which it occurs:

Storm Type	Region
I Typhoon	I Indian Ocean
I Hurricane	I Northwest Pacific
I Cyclone	I North Atlantic and East Pacific

5 Match each country or region to its dominant climatic region.

Country	Climatic Region
I Finland	I Desert
I Greece	I Humid subtropical
I New Zealand	I Mediterranean
I Oman	I Rainforest
I Uruguay	I Subarctic
I Indonesia	I Temperate

Mount Rainier

Tennessee-Alabama-Ohio 2002

East Coast 2004, 2012

Mount Washing

Oklahoma City

S. Carolina-Virginia

Yuma

Louisiana 2005, 2008

Mount Waialeale

Texas 2008

Florida-Alabama 2004, 2005, 2008

W. Mexico 2002, 2004, 2009, 2011, 2014

Bahamas-E. USA 2004, 2005

Caribbean 200 2005, 2008, 2011, 2014

S. Mexico 2005

Central America 2005

N.E. Cari 2004, 20 2009, 20

Atacama Desert

Major Climatic Regions and Tropical Storms

- • Weather extreme location
- → Cyclone track
- → Typhoon track
- → Hurricane track
- ⦿ Major tropical storm (1994–2015)

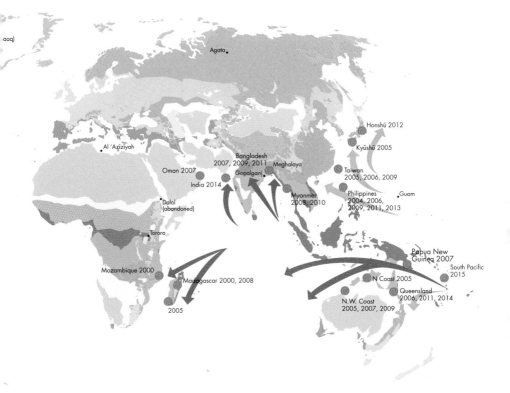

aaq)

Agata.

Honshū 2012

Kyūshū 2005

Al 'Aziziyah

Oman 2007

Bangladesh
2007, 2009, 2011   Meghalaya

Gopalganj

India 2014

Taiwan
2005, 2006, 2009

Dalol
(abandoned)

Myanmar
2008, 2010

Philippines
2004, 2006,
2009, 2011, 2013

Guam

Tororo

Papua New
Guinea 2007

South Pacific
2015

Mozambique 2000

Madagascar 2000, 2008

N Coast 2005

Queensland
2006, 2011, 2014

2005

N.W. Coast
2005, 2007, 2009

Polar	Ice cap
	Tundra

Cooler humid	Subarctic
	Continental cool summer
	Continental warm summer

Warmer humid	Temperate
	Humid subtropical
	Mediterranean

Dry	Steppe
	Desert

Tropical humid	Savanna
	Rainforest

2 CLIMATE GRAPHS

The graphs on the opposite page plot average temperature and precipitation for various worldwide cities that all have cool summers.

1 Which city has the highest average precipitation in August?

2 Which city has the lowest temperatures, on average, in January?

3 Which of the cities shown opposite has the highest elevation?

4 Which city, on average, does not receive more than 50mm of precipitation in any month of the year?

5 In Tomsk, what is the difference between the highest average temperature in July and the lowest average temperature in December?

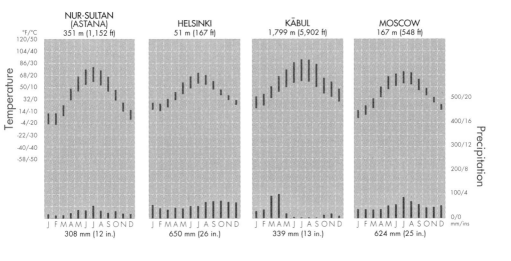

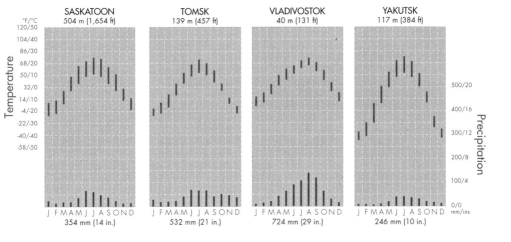

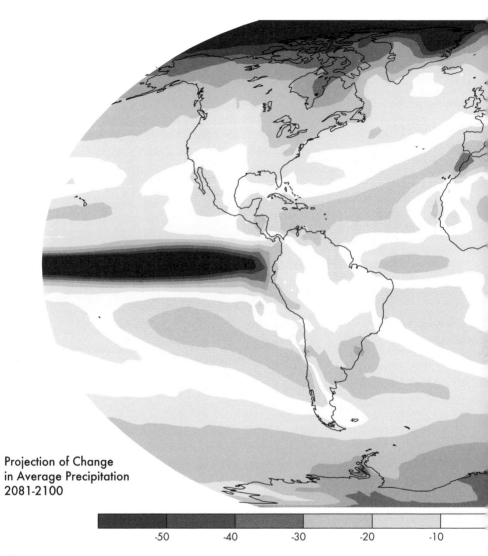

Projection of Change
in Average Precipitation
2081-2100

| -50 | -40 | -30 | -20 | -10 |

3 CHANGE IN PRECIPITATION

In the high-latitude regions (central and northern Europe, Asia and North America) the year-round average precipitation is projected to increase over the coming century, while in most sub-tropical land regions it is projected to decrease by as much as 20 per cent. This would increase the risk of drought and, in combination with higher temperatures, threaten agricultural productivity.

1 Which continent will see the least overall change in its precipitation?

2 Which country will see the greatest drop in its precipitation towards the end of the century?

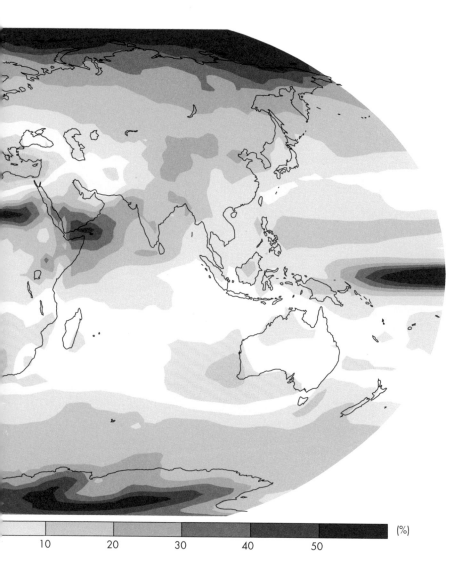

(%)

10 20 30 40 50

3 Which continent will see the greatest increase in precipitation?

4 Which continent is predicted to see the greatest variation in its changes in precipitation towards the end of the century?

4 WEATHER EXTREMES

1 On the opposite page is a list of extreme weather records as of 2019, and the places where these records occurred.

Match the data values to the records they correspond with.

Data values:

Temperatures:

    | -56.6°C
    | 34.4°C
    | 56.7°C

Lengths:

    | 0.1mm
    | 11.873m
    | 31.102m

Time periods:

    | up to 350 per year
    | over 4,000 hours
    | 251 days per year
    | nil for 182 days each year

RECORD	LOCATION	VALUE
Highest shade temperature	Furnace Creek, Death Valley, California, USA	
Hottest place (annual mean)	Dalol, Ethiopia	
Driest place (annual mean)	Atacama Desert, Chile	
Most sunshine (annual mean)	Yuma, Arizona, USA	
Least sunshine	South Pole	
Coldest place (annual mean)	Plateau Station, Antarctica	
Wettest place (annual mean)	Meghalaya, India	
Most rainy days	Mount Waialeale, Hawaii	
Greatest snowfall	Mount Rainier, Washington, USA	
Thunder-days average	Tororo, Uganda	

5 WORLD POPULATION DISTRIBUTION

This map shows the regions of the world that have the most and the least people living in them. Using the information on the map, answer the following questions.

1 Which part of the world has the greatest area with a population in excess of 2,500 people per square mile?

2 Not including Antarctica, what is the least densely populated territory in the world?

3 Of all the continents, which is the most sparsely populated?

4 What is the most densely populated island in Indonesia?

5 Rank the following regions from highest (1) to lowest (6) in terms of total population.

| Africa
| Europe (including Russia)
| Oceania
| Asia
| Northern America
| South and Central America, and the Caribbean

World Population Distribution

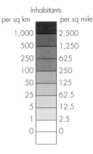

Inhabitants

per sq km		per sq mile
1,000		2,500
500		1,250
250		625
100		250
50		125
25		62.5
5		12.5
1		2.5
0		0

6 POPULATION GROWTH

The world's population is steadily increasing, but the populations of individual continents are not all rising at the same rate. The graph below shows the rate of growth for each of the continents, both historically and as projected to 2050.

1 In which decade did the population of Africa overtake that of Europe?

2 What was the population of the world in 1800?

3 The population of which region is projected to rise above that of which other for the first time in modern history in around 2050?

4 In the year 2000, approximately how many more people were there in Asia than in Latin America and the Caribbean?

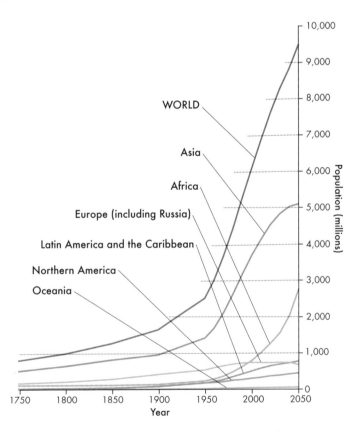

7 POPULATION QUIZ

1 China introduced a one-child policy in 1979, penalizing parents who had more than one child. When was the policy abolished?

2 Which continent currently has the highest average life expectancy?

3 Back in 1950, which was the only city with over 10 million inhabitants?

 I New York
 I New Delhi
 I Mexico City

4 Which is predicted to be the largest city in the world in 2020, in terms of its population?

 I São Paulo
 I Tōkyō
 I New Delhi

5 What was the first year in which half of the world's population were living in urban areas?

 I 2002
 I 2005
 I 2007

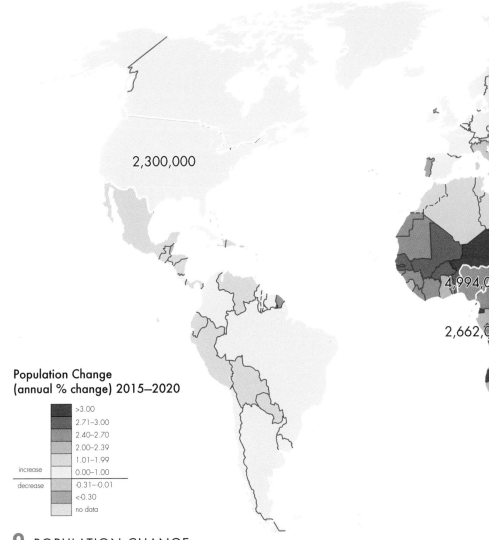

2,300,000

4,994,0

2,662,0

**Population Change
(annual % change) 2015–2020**

	>3.00
	2.71–3.00
	2.40–2.70
	2.00–2.39
	1.01–1.99
increase	0.00–1.00
decrease	-0.31–-0.01
	<-0.30
	no data

8 POPULATION CHANGE

Just as the populations of the continents grow at different rates, so do the populations of individual countries. This map shows the average annual rate of population change for each country of the world, shown as a percentage of the total population. The net annual additions of the top ten contributors to population change are indicated, and those countries are outlined on the map.

1 Is the population of Russia currently increasing or decreasing?

2 Which countries are the top ten contributors to global population growth? They can be found outlined in white on the map, and their net annual increases are indicated.

3 Below is a list of capital cities belonging to countries which all have the same rate of population change.

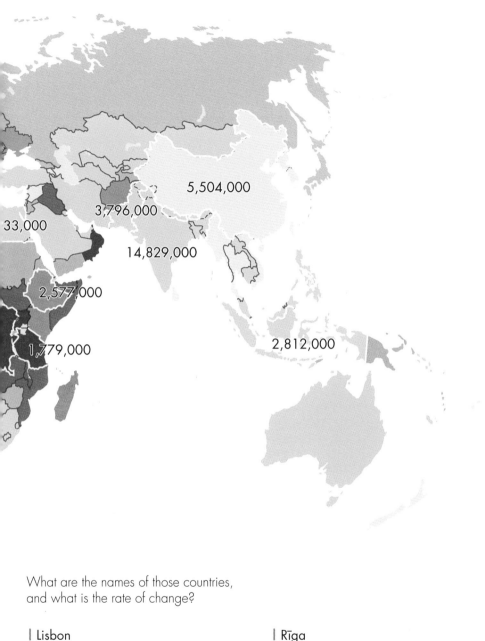

5,504,000

3,796,000

33,000

14,829,000

2,577,000

1,779,000

2,812,000

What are the names of those countries,
and what is the rate of change?

| Lisbon

| Rīga

| Kiev

| Sofia

| Bucharest

9 CITY POPULATIONS

1 Complete this list of the top 5 most populated cities on each continent. The first letter of each city name is given, but remember that some names may contain more than one word. When you are done, order all of the cities within each continent by population, from 1 (highest population) to 5 (lowest population).

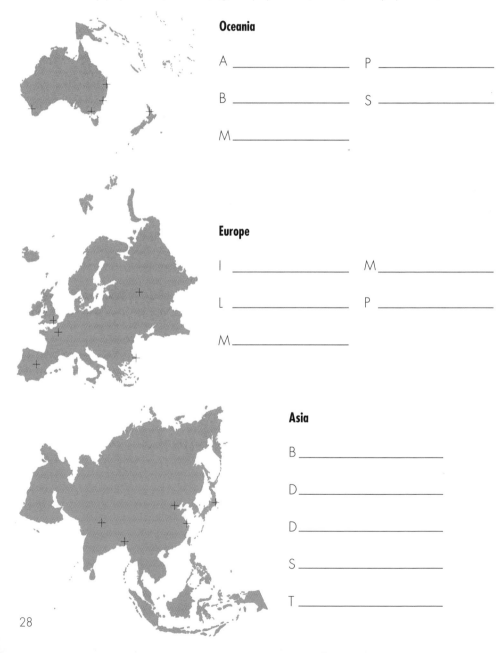

Oceania

A _____   P _____

B _____   S _____

M _____

Europe

I _____   M _____

L _____   P _____

M _____

Asia

B _____

D _____

D _____

S _____

T _____

South America

B _____

B _____

L _____

R _____

S _____

Africa

C _____

D _____

K _____

L _____

L _____

North America

C _____

H _____

L _____

M _____

N _____

10 MOST POPULOUS CITIES

The map opposite shows those cities of the world which are projected to have a population in excess of 1 million by 2020. Those with populations in excess of 5 million are also named.

1 How many cities in South America are projected to have more than 10 million inhabitants?

2 Which continent has the greatest number of people living in cities?

3 Of those cities with more than 20 million inhabitants, how many are also capital cities?

4 Which country, one of the ten largest in the world by area, has six cities with a population of more than 1 million inhabitants, but none with a population of more than 10 million inhabitants?

5 Can you name two African countries, both south of Kinshasa, which do not have any cities with more than 1 million people living in them?

The World's Major Cities 2020

Number of inhabitants
(cities with over 5 million are named)

- 1 million–2.5 million
- 2.5 million–5 million
- 5 million–10 million
- 10 million–20 million
- over 20 million

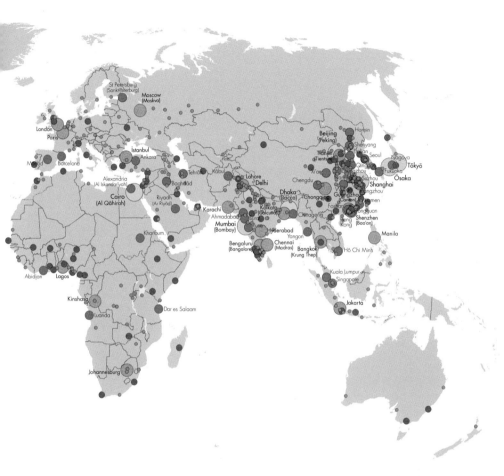

Figures are for the urban agglomeration, defined as the population contained within the contours of a contiguous territory inhabited at urban levels without regard to administrative boundaries. They incorporate the population within a city plus the suburban fringe lying outside of, but adjacent to, the city boundaries.

11 CITY CLOSE-UPS

On the next three pages are views of six city centres.

1   Which is the only one of these city centres not to use roundabouts or traffic circles in their street layout?

2   Most of these city centres have multiple museums:

  ▎How many city centres have what is labelled as a 'National Museum'?

  ▎Which city centres have museums named after the city itself?

  ▎How many of these cities have a Natural History Museum shown?

3   How many of these cities have at least one university labelled? Which cities have two?

4   How many different religions can you find places of worship for on the maps? Which religion is represented in the greatest number of these city centres?

5   Which location sounds like a well-reviewed place to wash some clothes?

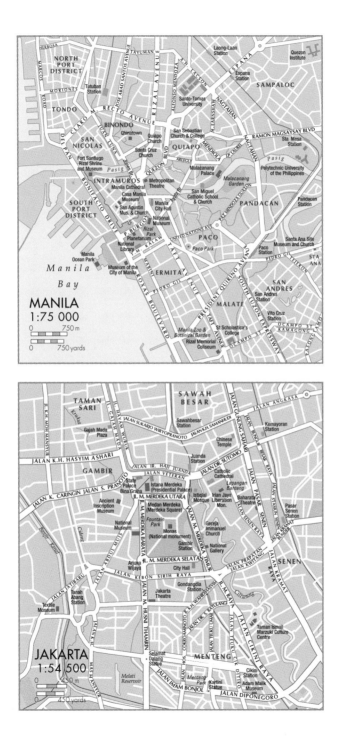

MANILA
1:75 000

0   750 m

0   750 yards

Manila
Bay

JAKARTA
1:54 500

0   450 m

0   450 yards

SHANGHAI
1:75 000

0 — 750 m
0 — 750 yards

JINGAN

Shanghai Natural History Museum
Jing'an Sculpture Park
BEIJING XILU
Jing'an Temple
NANJING XILU
Jing'an Park
Shanghai Exhibition Centre
Rujin Theatre
YAN'AN
Lyceum Theatre
Indigo Batik Museum
Xiang Yang Park
HUAIHAI
Conservatory of Music
Former Residence of Sun Yat-Sen
Fuxing Park
FUXING ZHONGLU
Cultural Square
Former Residence of Zhou En-Lai
CHANGSHU LU
ZHONGLU
YAN'AN
Acrobatic Theatre
Grand Theatre
People's Square
Shanghai Museum
Dazhong Theatres
Huaihai Park
Site of the First National Congress of the Chinese Communist Party
LUWAN
Hunan Stadium
XIELU JIANGUO
RUJIN DONGLU JIANGUO
XIUHUAI
LUJIABANG LU

TIANMU LU
HENGFENG LU
Wusong River
CHENGDU
MENGZI
ANQING LU
Shanghai Railway Museum
Jade Buddha Temple
BEIJING DONGLU
No. 1 Department Store
Friendship Store
NANJING DONGLU
Renmin (People's) Park
Muen Church
Worker's Cultural Palace
Great World Entertainment Centre
RENMIN
Yuyuan Garden
NANSHI
Confucian Temple
HONGQIAO LU
Penglai Park
ZHONGSHAN NAN ZHONGSHAN
CHEZHAN QIANLU
NAN LU

DONGDAMING LU
Shanghai People's Hero Memorial Pagoda
The Bund Historical Museum
Shanghai International Conference Centre
Pearl of the Orient TV Tower
Aurora Art Museum
The Bund
Yan'an Donglu Tunnel
FUXING DONGLU
DONGLU

HUANGPU
Huangpu Jiang
PUDONG
ZHONGSHAN DONG ZHU
Nanpu Bridge
PUDONG
ZHONGSHAN

PUDONG NEW AREA
Huangpu Jiang

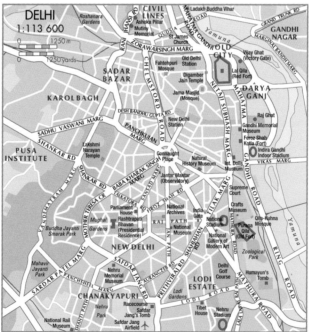

DELHI
1:113 600

0 — 1250 m
0 — 1250 yards

Roshanara Gardens
CIVIL LINES
Ashoka Pillar
Mutiny Memorial
QUDSIA
St James Church
RANI HANSLO
ZORAWARSINGH MARG
SADAR BAZAR
Fahtehpuri Mosque
Old Delhi Station
Digamber Jain Temple
Jama Masjid (Mosque)
KAROLBAGH
DESH BANDHU GUPTA RD
New Delhi Station
SADHU VASWANI MARG
PANCHKUIAN MARG
SHANKAR RD
PUSA INSTITUTE
Lakshmi Narayan Temple
Connaught Place
VANDEMATRAM MARG
Buddha Jayanti Smarak Park
MOTHER TERESA CR
SHANKAR RD
BABA KHARAK SINGH MARG
Jantar Mantar (Observatory)
KAIKAYE
Parliament House
Rashtrapati Bhavan (Presidential Residence)
Mughal Gardens
RAJ PATH
National Museum
NEW DELHI
Mahavir Jayanti Park
SARDAR PATEL MARG
Nehru Memorial Museum
PANCHSHEEL MARG
SAFDARJANG RD
CHANAKYAPURI
SHANTI PATH
Nehru Park
Racecourse
Safdar Jang's Tomb
Safdar Jang Airfield

Ladakh Buddha Vihar
MAHATMA GANDHI ROAD
Yamuna
GRAND TRUNK RD
MARGINAL BANDH MARG
GANDHI NAGAR
OLD CITY
Vijay Ghat (Victory Gate)
Lal Qila (Red Fort)
NETAJI SUBHASH MARG
DARYA GANJ
Raj Ghat
Gandhi Memorial Museum
Feroz Shah Kotla (Fort)
MAHATMA
Indira Gandhi Indoor Stadium
VIKAS MARG
Int. Dolls Museum
Supreme Court
TILAK MARG
India Gate
National Stadium
National Gallery of Modern Art
Crafts Museum
Qila-Kuhna Mosque
Purana Qila (Old Fort)
Yamuna
Zoological Park
Delhi Golf Course
Humayun's Tomb
LODI ESTATE
Lodi Gardens
Tibet House
Nehru Stadium
RING ROAD
MATHURA ROAD
LODI ROAD
National Rail Museum

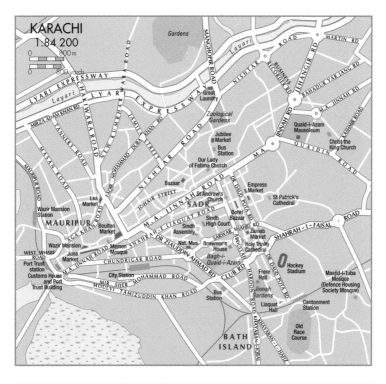

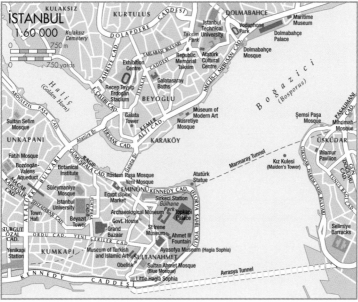

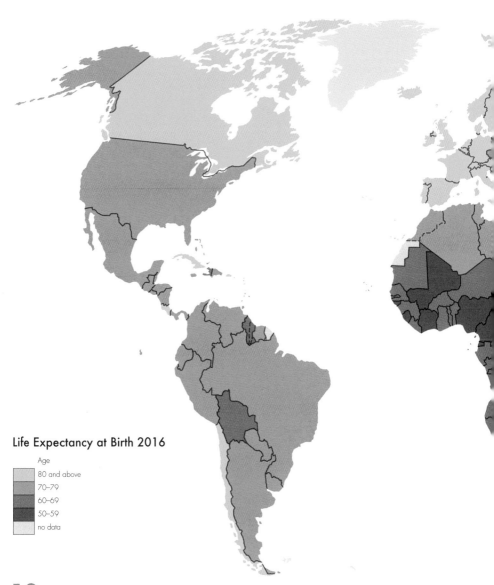

Life Expectancy at Birth 2016

Age

- 80 and above
- 70–79
- 60–69
- 50–59
- no data

12 LIFE EXPECTANCY

Thanks to worldwide improvements in medicine and healthcare, the average life expectancy for a human being has increased dramatically in the last two centuries. The map on this page reveals the average life expectancy at birth for every country of the world.

1 The world's two largest economies have lower life expectancies than many other countries. Which two countries are these?

2 Which large, landlocked country with a life expectancy of 60–69 is bordered only by countries with a life expectancy of 70–79?

3 Which is the only country in South America with a life expectancy of 80+? And which two South American countries share the lowest life expectancy range on the continent?

4 What is the median life expectancy range in Africa? (The median is the middle value in a sorted list).

5 Two Asian locations share the world's greatest life expectancy of 84 years. Which country and territory do you think these are?

13 DEATHS FROM POOR HEALTHCARE

In spite of major healthcare developments, some countries still show a relatively high percentage of deaths caused by communicable diseases and maternal, prenatal and nutritional conditions. The map shows what percentage of the population dies from these factors in each country.

1 Which African country with a death rate of 60% and above is bordered by the greatest number of other countries with an equally poor rate?

2 Which two countries on the Arabian Peninsula have the highest percentage of deaths caused by communicable diseases and, maternal, prenatal and nutrition conditions?

3 Which country has the greatest difference in its healthcare quality compared to that of a pair of neighbouring countries?

4 By also using the map on the previous page, can you find two countries with a death rate, as shown on this map, of 14.9% or lower and yet with a life expectancy of less than 70?

5 Although this particular distinction cannot be seen in the ranges marked on the map, which country do you think has the least effective healthcare outside of Africa?

Cause of Death: Communicable Diseases and Maternal, Prenatal and Nutrition Conditions 2015

% of population

- 60 and above
- 45–59.9
- 30–44.9
- 15–29.9
- 0–14.9
- no data

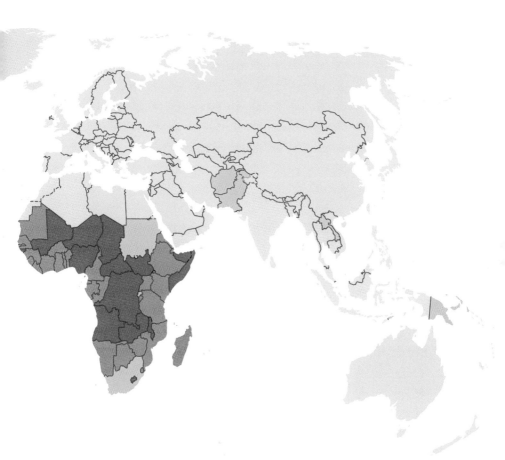

14 HEALTH QUIZ

1 Since 1800, how has average global life expectancy changed?

I It has increased from 30 years to 70 years.

I It has increased from 25 years to 65 years.

I It has increased from 20 years to 65 years.

2 How many worldwide deaths each year can be attributed to a combination of poor diet and tobacco use?

I Approximately a quarter.

I Approximately a third.

I Approximately a half.

3 Child mortality rates in Africa were how many times higher than in Europe, according to 2016 data?

I 2 times

I 4 times

I 8 times

4 Many people live with chronic, non-fatal illnesses or disabilities, which are not captured by death statistics. To estimate the overall effects of such quality of life issues, a measure called the DALY is used, which combines a measure of both early death and years of lower-quality life. What does DALY stand for?

I Daily Average Life.

I Disability Adjusted Life Year.

I Disease Adapted Life Year.

5 In what year did the World Health Organization (WHO) declare smallpox eradicated from the world?

I 1970

I 1980

I 1990

6 According to the WHO, what was the percentage drop in worldwide deaths from malaria during the period from 2000 to 2015?

| 25%

| 48%

| 60%

7 And which two countries were certified malaria-free by the WHO in 2019?

| Argentina and Algeria

| Bolivia and Botswana

| Chile and Chad

8 Which country was rated the happiest in the 2019 UN World Happiness Report?

| Fiji

| Finland

| France

9 Social resistance to what critical method of disease prevention has recently led to many deaths from illnesses that had previously been considered controlled at low levels?

| Isolation

| GP visits

| Vaccination

10 Growing resistance to what type of medicine has made many drugs become ineffective, sometimes resulting in illnesses that can no longer be cured?

| Retroviral

| Antibiotic

| Immunosuppressant

15 TRADE

The map opposite reveals how countries compare in terms of their level of international trade, scaled as a percentage of their domestic earnings.

1 Which is the most southerly country on the map with trade equal to more than 125% of its gross domestic product (GDP)?

2 How many countries in mainland Africa have trade equal to between 75 and 100% of their GDP?

3 Out of the five largest countries in the world by area – Russia, Canada, USA, China and Brazil – which has the highest level of trade, relative to its GDP?

4 Based on the map, which continent has the lowest level of trade, relative to its GDP?

5 Although not specifically indicated on the map, which multilingual EU country do you think has the highest level of trade of any country, relative to its GDP?

Trade as a Percentage of GDP 2016

125
100.1–124.9
75.1–100
50.1–75
0–50
no data

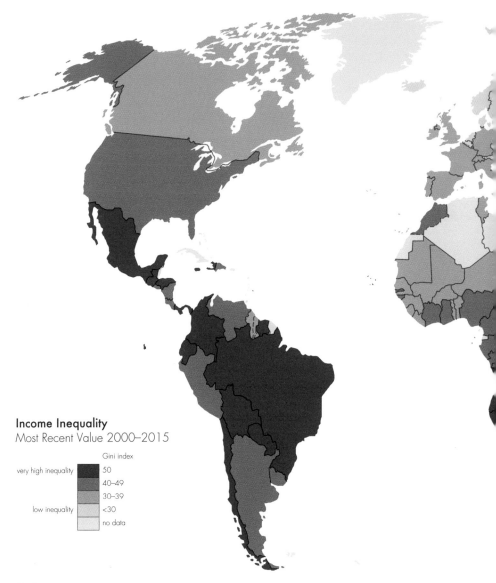

Income Inequality
Most Recent Value 2000–2015

Gini index

very high inequality	50
	40–49
	30–39
low inequality	<30
	no data

16 INCOME INEQUALITY

The map opposite reveals varying levels of income inequality between countries, with darker colours indicating a greater gap between the highest and lowest earnings. These are normalized using a measure known as the Gini index, where 0 represents perfect equality (all people earning the same) and 100 represents

perfect inequality (all money earned by a single individual).

1 Which is the largest country by area to be ranked as having 'very high' income inequality?

2 On which two continents does income

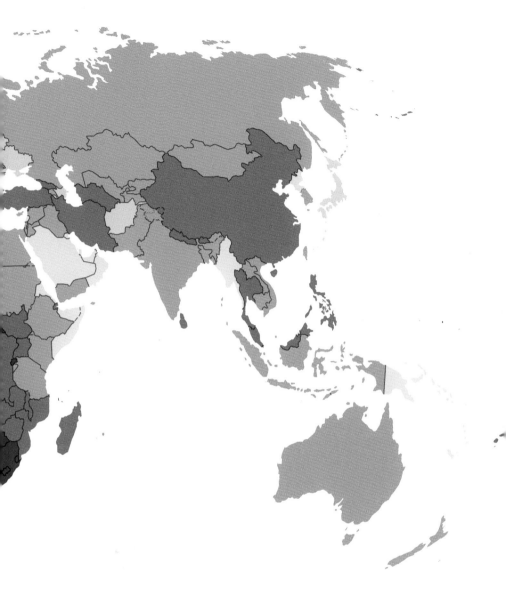

inequality broadly increase as you head from north to south?

the lowest overall level of income inequality?

3 Which three Asian countries have the lowest income inequality? Exclude those for which no data is available.

4 Excluding Oceania, which of the continents shown on the map has

5 By comparing with the map on the previous page, can you find five countries with trade equal to at least 100% of their GDP, but also 'very high' income inequality?

17 GROSS NATIONAL INCOME

The map opposite shows how gross national income varies by country, normalised both by population size and by relative purchasing power of each country's currency. This therefore provides a measure of the average earnings of an individual per country in a way that can be directly compared across the world.

1 Two of the five countries with the highest average income per capita are located in east Asia. Which two countries are they?

2 And in which other region of the world do you think the three other countries in the top five average income per capita are located?

3 Based on this map, which country would you expect to have the highest total national income in the world?

4 Which country is most likely to have the highest median salary? You can deduce this by combining information from this and the income inequality chart in this section.

5 There are three European countries with GNI per capita in excess of $60,000. Which of these do you think has the highest average income?

Gross National Income (GNI) Per Capita 2016

GNI per capita, purchasing power parity (PPP) (current international $)

▓	>60 000
▓	30 000–60 000
▓	10 000–29 999
▓	3 000–9 999
▓	0–2 999
▓	no data

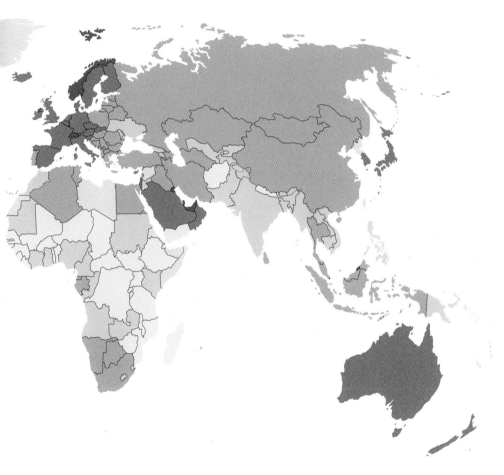

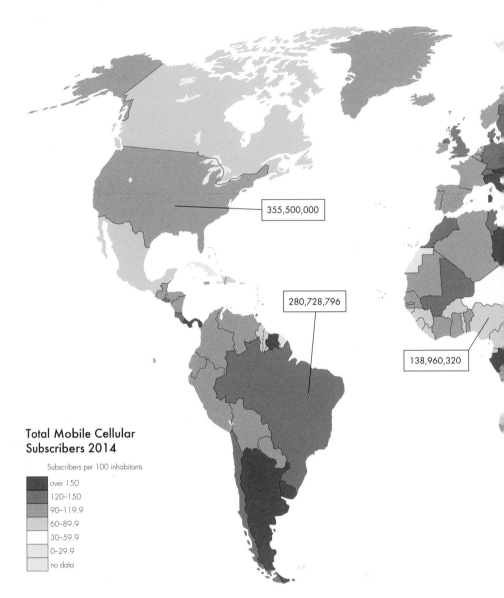

355,500,000

280,728,796

138,960,320

**Total Mobile Cellular
Subscribers 2014**

Subscribers per 100 inhabitants

	over 150
	120–150
	90–119.9
	60–89.9
	30–59.9
	0–29.9
	no data

18 MOBILE PHONE USERS

Mobile phone ownership is by no means equal across the world. This map shows the number of mobile phone subscriptions per 100 inhabitants of each country. Numbers above 100 indicate that, on average, each person in a country owns more than one phone.

1 Of the ten countries with the most subscribers, which has the highest density of cellular phones per person?

2 In which three African countries does every person own, on average, at least one and a half phones? Their initial letters are B, G and L.

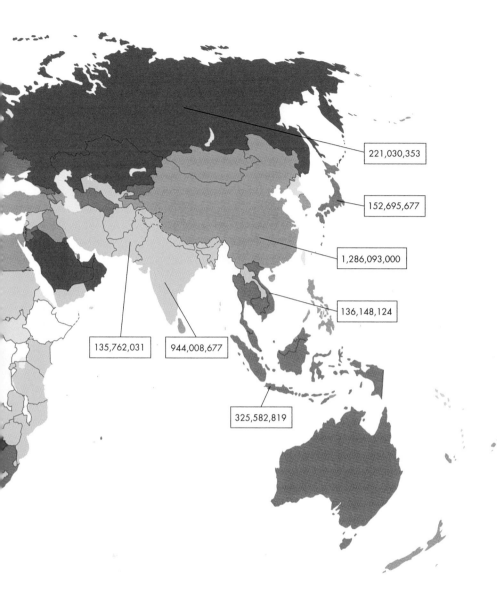

3 Using just the map and its data, can you calculate lower and upper bounds on the total population of the USA, as of 2014?

4 Name two countries shown on the map as having no cellular subscriber information?

5 Which recently formed country has the joint-lowest density of cellular phone usage, compared to all other countries for whom data is provided?

19 TELECOMMUNICATIONS

The upper graph on the opposite page shows the change in the number of people worldwide using different communication methods over recent years. The graphs below reveal the varying usage of communication methods per continent.

1 In what year did the number of internet users overtake the number of main telephone lines?

2 By approximately what percentage did the number of mobile cellular subscribers increase between 2004 and 2008?

3 When did the total number of mobile cellular subscribers surpass more than half of the global population count?

4 As of 2014, how many main telephone lines were there in Africa?

5 Which were there more of in 2014: broadband subscribers in the Americas, or mobile cellular subscribers in Africa?

World Communications Equipment

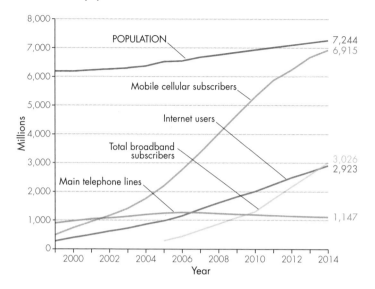

International Telecommunications Indicators

Main telephone lines

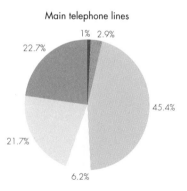

1%   2.9%
22.7%
45.4%
21.7%
6.2%

Fixed broadband subscribers

0.5%   1.7%
23.1%
44.4%
24.6%
5.7%

Mobile cellular subscribers

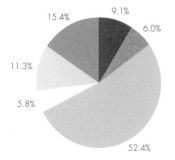

9.1%
15.4%
6.0%
11.3%
5.8%
52.4%

- Africa
- Middle East
- Asia and Pacific
- Former Soviet Union
- Europe
- The Americas

20 COMMUNICATIONS QUIZ

1 Ten record-breaking feats of human construction are represented by the lines on the map. Match each construction to its correct location.

2 In what year was the first submarine transatlantic telephone cable laid, running between Scotland and Newfoundland?

3 Which is the only continent not connected to the world-wide fibre-optic network?

4 To the nearest tens of thousands, how many text messages were sent globally every second in 2010?

5 As of 2016, what percentage of the world's internet servers were located in North America?

6 Wikipedia has more articles about people and places on which continent, relative to all the other continents combined?

7 Africa has 11% of the world's population, but what percentage do you think it has of the world's domain name registrations?

a _____

i _____

i _____

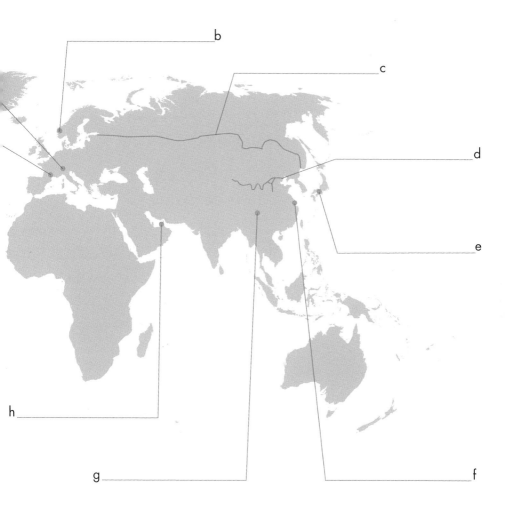

b

c

d

e

h

g

f

| Longest railway tunnel: Gotthard Base Tunnel (57 km/35.5 miles)

| Longest suspension bridge: Akashi-kaikyō Bridge (1,991 m/6,532 ft)

| Tallest building: Burj Khalifa (828 m/2,717 ft)

| Longest bridge: Danyang-Kunshan Grand Bridge (164 km/102 miles)

| Longest road tunnel: Laerdal Tunnel (24.51 km/15.23 miles)

| Longest railway: Trans-Siberian Railway (9,289 km/5,722 miles)

| Tallest dam: Jinping-I Dam (305 m/1,001 ft)

| Longest road: Pan-American Highway (30,000 km/19,000 miles)

| Tallest road bridge: Millau Viaduct (343 m/1,125 ft)

| Longest wall: Great Wall of China (21,196 km/13,171 miles)

21 WORLD TIME ZONES

For the questions below, assume that daylight saving time is not being taken into account. Note that the time zone that spans the Prime Meridian is known as Coordinated Universal Time (or UTC), and is equivalent to Greenwich Mean Time (GMT).

1 If the time in the UK is 15:30, what is the time on the east coast of Mexico?

2 Which country spans five geographical time zones, yet uses only one?

3 If you flew from Helsinki in Finland, leaving at 11a.m. on Friday, on a 22-hour flight that took you to Port Moresby in Papua New Guinea, what would the local time and day be when you land?

4 Which country has the greatest number of different time zones, excluding time zones that are only used for overseas territories of a country?

5 Can you match up these cities into pairs that share a time zone?

| Addis Ababa, Ethiopia
| Beijing, China
| Ottawa, Canada
| Dakar, Senegal
| Paris, France
| Johannesburg, South Africa

| Lagos, Nigeria
| Moscow, Russia
| Reykjavík, Iceland
| Lima, Peru
| Perth, Australia
| Helsinki, Finland

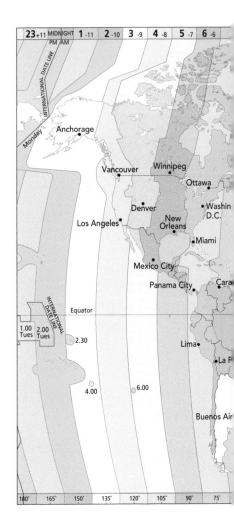

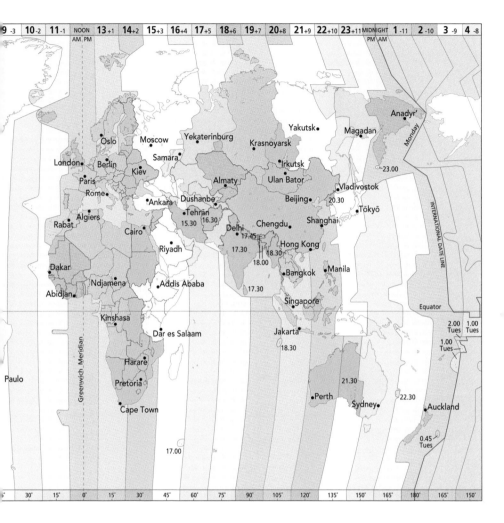

Anadyr'

Monday

Yakutsk

Magadan

Oslo
Moscow
Yekaterinburg
Krasnoyarsk
23.00

London
Berlin
Kiev
Samara
Irkutsk

Paris
Rome
Almaty
Ulan Bator
Vladivostok
20.30

Ankara
Dushanbe
Beijing

Rabat
Algiers
Tehrān
15.30
16.30
Chengdu
Shanghai
Tōkyō

Cairo
Delhi
17.45

Riyadh
17.30
18.30
Hong Kong
18.00

Dakar

Ndjamena
Addis Ababa
Bangkok
Manila
17.30

Abidjan

Singapore
Equator

Kinshasa

Dar es Salaam
Jakarta
18.30
2.00 Tues
1.00 Tues

1.00 Tues

Harare

Pretoria
21.30

Cape Town
Perth
22.30
Sydney
Auckland

Paulo

Greenwich Meridian

INTERNATIONAL DATE LINE

17.00

0.45 Tues

22 SUMMER AND WINTER TIMES

The table below shows some time zones that have daylight-saving variations, when an extra UTC +1:00 is added to the standard time. The months during at least part of which daylight-saving time is in effect are also shown. Use this information to answer the following questions.

1 Is the time difference between Paraguay and Phoenix, Arizona (which uses Mountain Time) larger, smaller, or the same when comparing between April and December?

2 When the time is 6:45 p.m. ACDT, what time is it in IST?

3 If it is 12 p.m. in July in Phoenix, Arizona, what time is it in Adelaide (which uses Australian Central Time)?

4 If it's 3:30 a.m. in Israel in September, what time is it in Paraguay?

5 When it's midnight on Sunday in Fiji in December, what time and day is it in Israel?

Standard time shift	Standard time zone	Daylight-saving time zone	Months daylight-saving in effect
UTC+9:30	ACST – Australian Central Standard Time	ACDT – Australian Central Daylight Time	October to April
UTC-7	MST – Mountain Standard Time	MDT – Mountain Daylight Time	March to November
UTC+12	FJT – Fiji Time	FJST – Fiji Summer Time	November to January
UTC+2	IST – Israel Standard Time	IDT – Israel Daylight Time	March to October
UTC-4	PYT – Paraguay Time	PYST – Paraguay Summer Time	October to March

23 TIME ZONE ABBREVIATIONS

How many of the following time zones can you identify, from their standard abbreviations? Which country or region do they belong to?

| ADT

| AEST

| CET

| CST

| CXT

| EAT

| ECT

| EET

| HKT

| IDT

| JST

| MSK

| NT

| NZDT

| SST

| WAT

24 THE UNITED STATES

The US spans several different time zones, as shown by the map opposite.

1 Can you list the names of the four mainland time zones in the United States?

2 How many of the five other official US time zones can you name?

3 Only six of these time zones observe changes for daylight-saving time. Which three do not?

4 The USA maintains three Antarctic research stations. What is the time zone at Palmer Station, located on Anvers Island in the Antarctic Peninsula region?

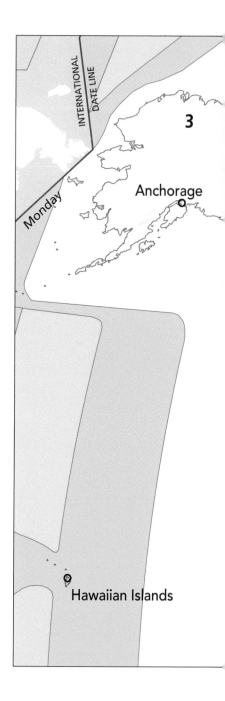

4

5

6

7

8

Denver

Washington
D.C.

Los
ngeles

New
Orleans

Miami

Puerto Rico

US Virgin Islands

59

GEOGRAPHICAL INFORMATION

25 WORLD'S HIGHEST MOUNTAINS

The illustration shows the ten highest mountains in the world. Their names are given below, but they have been encoded.

Crack the codes to reveal the original names.

For codes a to f, a Caesar shift has been applied. This means that each letter has been shifted by a constant amount, so for example with a shift of 1 then 'a' would be changed to 'b', 'b' would be changed to 'c', and so on, wrapping around with 'z' being changed to 'a'. Two different shifts have been used, across the six mountain names. For codes g to j, it's up to you to work out what is going on, but again two different encodings are used.

a. NPVOU FWFSFTU

b. DIPHPSJ GFOH

c. NDQJFKHQMXQJD

d. OKRWVH

e. PDNDOX

f. FKR RBX

g. 4-8-1-21-12-1-7-9-18-9-0-9

h. 13-1-14-1-19-12-21

i. 13-26-13-20-26-0-11-26-9-25-26-7

j. 26-13-13-26-11-6-9-13-26-0-18

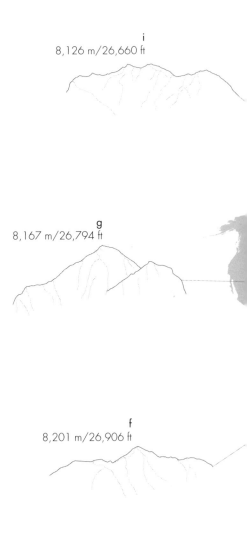

i
8,126 m/26,660 ft

g
8,167 m/26,794 ft

f
8,201 m/26,906 ft

62

b
8,611 m/28,251 ft

i
8,091 m/26,545 ft

h
8,163 m/26,781 ft

a
8,848 m/29,028 ft

c
8,586 m/28,169 ft

d
8,516 m/27,939 ft

e
8,463 m/27,765 ft

26 MOUNTAIN QUIZ

1   How many of the top ten highest mountains in the world straddle the border of China and Nepal?

2   True or false: all of the top forty highest mountains in the world are in Asia?

3   Which two climbers are recorded as being the first to reach the summit of Everest, in 1953?

4   How many countries do the Himalayas span?

5   If you include height below sea level for mountains that emerge from the sea, Everest is not the world's highest mountain. What is it, and in which country is it located?

6 What are, respectively, the names of the highest mountains in Scotland, Wales and England?

7 Which South American mountain chain is the longest in the world, excluding those below sea level?

8 Which North American mountain system contains the Great Smoky Mountains, which in itself is a subrange of the Blue Ridge Mountains?

9 What is the height, in metres, of the start of the altitude zone in mountaineering in which it is generally considered that there is not enough oxygen to sustain human life?

10 What unofficial boundary is partly formed by the Ural Mountains in Russia?

1 Match each country to its flag.

Barbados	Greece	Kenya
Mexico	New Zealand	Norway
Slovakia	South Africa	Spain

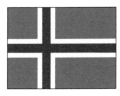  

  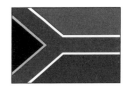

2 What does the circular symbol in the centre of the Japanese flag represent?

3 How many stars are there on the Chinese flag?

4 Which three countries do the outlines of these unusually shaped flags represent?

5 What feature do the flags of Iraq and Saudi Arabia both have in common, ignoring shared colours?

6 Which two countries have flags which feature two colours divided by a zigzag line?

7 If you reverse the colours of the Indonesian flag, what European country's flag does it become?

8 Which Scandinavian country is represented by each of the following flags?

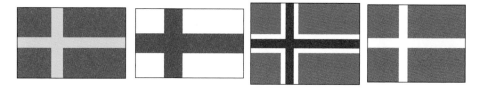

28 ISLAND MATCHING

Outlines of the world's ten largest islands are shown on the opposite page. Their names are given below.

Match each name to its corresponding island outline.

| Baffin Island
| Borneo
| Ellesmere Island
| Great Britain
| Greenland
| Honshū
| Madagascar
| New Guinea
| Sumatra (Sumatera)
| Victoria Island

a

b

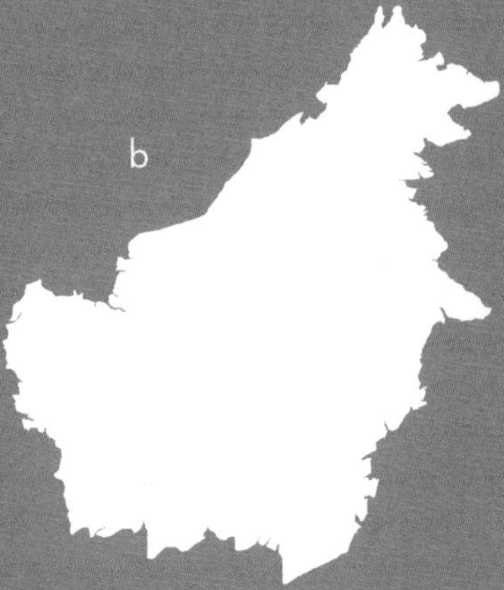

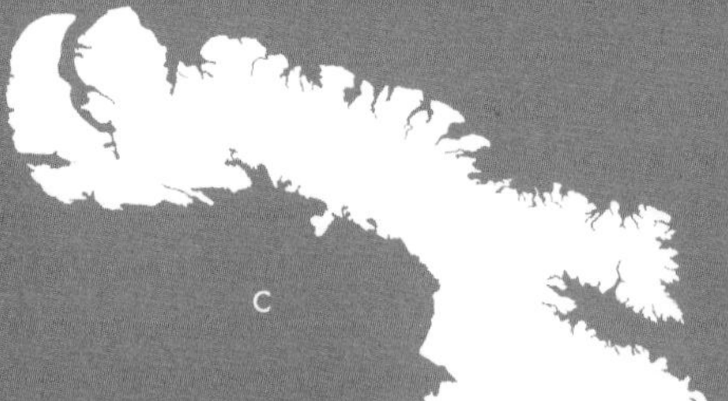

c

d

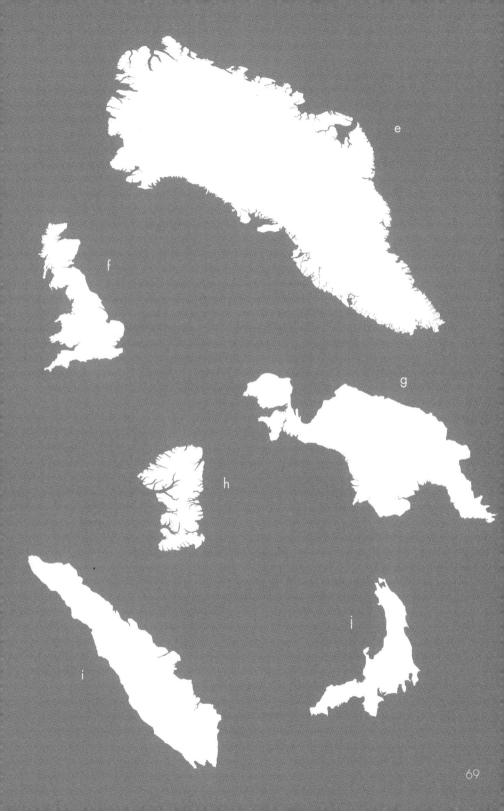

e

f

g

h

i

i

29 ISLAND QUIZ

1 Which country is Greenland a dependency of?

| Russia

| Canada

| Denmark

2 How many of the top ten largest islands in the world are part of Canada?

| 1

| 2

| 3

3 Which island is shared by three independent countries?

| Victoria Island

| Honshū

| Borneo

4 Which two countries share the island of New Guinea?

| Australia and Papua New Guinea

| Papua New Guinea and Indonesia

| Australia and New Zealand

5 Which of the following best describes the island of Honshū in relation to the four main islands that make up Japan?

| The most northerly

| The second most northerly

| The second most southerly

30 WORLD'S LARGEST ISLANDS

The names of the ten largest islands are given again below, along with a list of their areas - but these are not in the correct order.

Match the name of each island to its area.

| Baffin Island | 196,236 sq km

| Borneo | 217,291 sq km

| Ellesmere Island | 218,476 sq km

| Great Britain | 227,414 sq km

| Greenland | 473,606 sq km

| Honshu | 507,451 sq km

| Madagascar | 587,040 sq km

| New Guinea | 745,561 sq km

| Sumatra (Sumatera) | 808,510 sq km

| Victoria Island | 2,175,600 sq km

Great Bear Lake

Lake Victoria

Great Slave Lake

Lake Superior

Lake Huron

Lake Michigan

Lake Tanganyika

31 WORLD'S LARGEST LAKES

The illustration opposite shows the outlines, locations and names of the ten largest lakes in the world.

These lakes are so large that many of them neighbour multiple countries. Find the full list of countries neighboured by each lake. The number of countries adjoining each lake is given as a hint.

Lake Baikal
(Ozero Baykal)

Caspian Sea

Lake Nyasa (Lake Malawi)

Caspian Sea	5	Lake Nyasa (Lake Malawi)	3
Great Bear Lake	1	Lake Superior	2
Great Slave Lake	1	Lake Tanganyika	4
Lake Huron	2	Lake Victoria	3
Lake Michigan	1	Lake Baikal (Ozero Baykal)	1

32 LAKES QUIZ

1 Which is the largest lake that feeds into the Nile?

2 How many lakes comprise the Great Lakes of North America, and what are their names?

3 Out of the world's ten largest lakes, how many are saltwater?

4 In which country can you find the deepest lake in the world?

5 And in which country can you find what is thought to be the world's oldest lake?

6 What is the name of the strait that connects Lake Michigan to Lake Huron?

33 LAKE AREAS

The names of the ten largest lakes are given again below, along with a list of their areas – but they are not aligned correctly.

Match the name of each lake to its correct area.

| Caspian Sea | | 28,568 sq km
| Great Bear Lake | | 29,500 sq km
| Great Slave Lake | | 30,500 sq km
| Lake Huron | | 31,328 sq km
| Lake Michigan | | 32,600 sq km
| Lake Nyasa (Lake Malawi) | | 57,800 sq km
| Lake Superior | | 59,600 sq km
| Lake Tanganyika | | 68,870 sq km
| Lake Victoria | | 82,100 sq km
| Lake Baikal (Ozero Baykal) | | 371,000 sq km

a

The first letters of the four countries shown
are given, but not necessarily in the correct order.

Which country does each area represent?

b

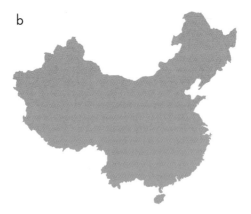

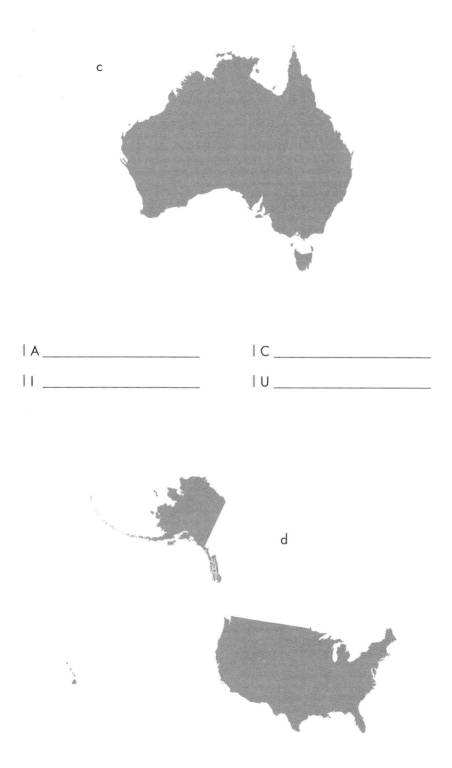

c

| A _____ | C _____ |
| I _____ | U _____ |

d

35 WORLD'S LONGEST RIVERS

The map opposite shows the paths and
locations of the world's ten longest rivers,
labelled a to j.

Identify which river corresponds with
each label.

| Amazon (Amazonas)

| Congo

| Mississippi-Missouri

| Nile

| Ob'

| Río de la Plata-Paraná

| Yangtze (Chang Jiang)

| Yellow River (Huang He)

| Yenisey-Angara-Selenga

| Irtysh (Yertis)

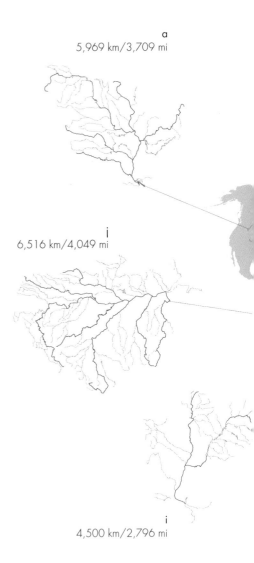

a
5,969 km/3,709 mi

i
6,516 km/4,049 mi

j
4,500 km/2,796 mi

b
,568 km/3,460 mi

c
5,550 km/3,449 mi

d
4,440 km/
2,759 mi

e
5,464 km/3,395 mi

f
6,380 km/3,965 mi

g
6,695 km/4,160 mi

h
,667 km/2,900 mi

36 RIVERS QUIZ

1 Which river flows through the greatest number of countries, compared to any other river in the world?

2 What is the name of the longest river in Europe?

3 Which river is fed by, among other sources, Lake Albert, Lake Tana and Lake Victoria?

4 On which river can you find the Three Gorges Dam?

5 Which river was responsible for forming the Grand Canyon?

37 THE MOUTH OF THE RIVER

For each sea listed below, which of the world's top-ten longest rivers drains into it? Beware though, more than one river drains into some of them.

I Bohai Sea

I East China Sea

I Gulf of Mexico

I Kara Sea

I Mediterranean Sea

I Central Atlantic Ocean

I South Atlantic Ocean

As a reminder, the rivers are as follows:

I Amazon (Amazonas)

I Congo

I Mississippi-Missouri

I Nile

I Ob'

I Río de la Plata-Paraná

I Yangtze (Chang Jiang)

I Yellow River (Huang He)

I Yenisey-Angara-Selenga

I Irtysh (Yertis)

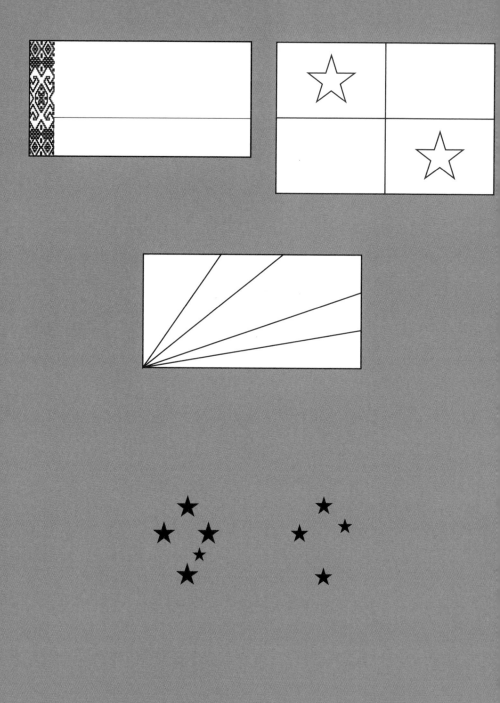

1 Identify which countries are represented by the first three flags on the opposite page, where all the colours have been removed.

2 These countries' flags all share a common feature, except for one. What is the shared feature, and which is the odd one out: Vietnam, Morocco, Bangladesh and Somalia?

3 Can you pair up the countries, one from each list, that have the same flag colours but in a different pattern, order or alignment?

Belgium	China
Canada	Colombia
Chad	France
Hungary	Germany
Ireland	Italy
Federated States of Micronesia	Côte d'Ivoire
Russia	Poland
North Macedonia	Somalia

4 On which five countries' flags can you find the Southern Cross constellation? This appears in either of the forms shown on the opposite page.

5 How many national flags feature either a rising or a setting sun crossing a horizon?

6 Which two flags feature an outline of the country they represent?

7 Which country depicts an AK-47 gun on its flag?

39 AFRICA AND EMPIRE

On the opposite page is a map of Africa, as of 1898. Many of the countries are shown as being overseas territories of European colonial powers.

Match the names of the ten 1898 territories listed below with the modern-day country or countries occupying roughly the same area today.

NAME (1898)	NAME (2019)
Abyssinia	Mozambique
British East Africa	Burundi, Rwanda and Tanzania
Cape Colony	Cameroon
Congo State	Democratic Republic of the Congo
French Guinea	Djibouti
French Somaliland	Ethiopia (part of)
German East Africa	Guinea
Kamerun	Kenya and Uganda
Portuguese East Africa	South Africa (southernmost part)
Rio de Oro	Western Sahara (western part of)

40 COUNTRY NAMES QUIZ

1 From 1922 to 1991, much of eastern Europe and northern Asia was part of the USSR. What does this acronym stand for? What was its Russian-language equivalent acronym?

2 What was the name of the Indian city of Chennai prior to its name change in 1996?

3 What is the country formerly known as Zaire now known as?

4 Which island did the Romans know as Hibernia?

5 Which modern country is the closest equivalent to the ancient country of Babylonia?

41 COUNTRY NAMES PUZZLES

1 Eleven countries have four-letter names. What are they?

2 What is the longest country name with alternating vowels and consonants?

3 Which letters of the alphabet are not the first letter of any country's name?

4 There are eighteen countries whose name begins with M. How many of them can you identify?

5 Which official country name (in English) has the greatest number of letters?

42 LARGEST COUNTRIES

The ten largest countries in the world by area are marked on the map, although their names have been replaced with letters from a to j.

Identify each country and then order them in size, from largest to smallest, by completing the table opposite.

Map letter	Country	Area (sq km)
		17,075,400
		9,984,670
		9,826,635
		9,606,802
		8,514,879
		7,692,024
		3,166,620
		2,766,889
		2,717,300
		2,381,741

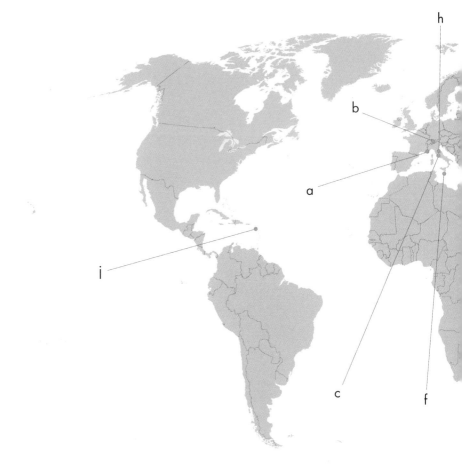

43 SMALLEST COUNTRIES

The ten smallest countries in the world are labelled on the map, but their names have been replaced with letters.

Complete the table on the opposite page by identifying the name of each of these countries, then place them in order of increasing size.

To help you, an obscured version of the name of each country is given. In each case, delete one letter from each pair of letters to reveal the name. For example, AF IK GJ EI would obscure FIJI: AF IK GJ EI. No spaces between words are marked, even if a country name is made up of multiple words.

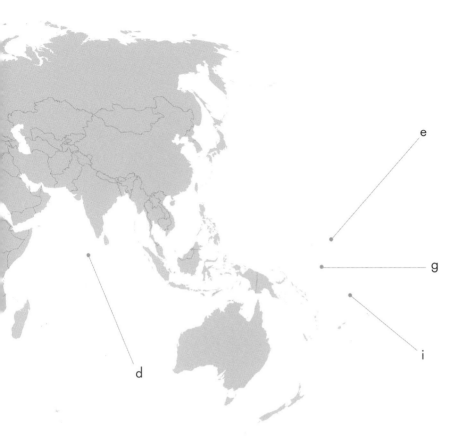

Map letter	Obscured country name	Country name	Area (sq km)
	VR PA IT LI CR EA MN NC ID UT YA		0.5
	LM IO NM AE EC NO		2
	NP AU UA LR DU		21
	TC EU AV PA BL EU		25
	DS LA MN AM AE RL IO GN HO		61
	LK IA LE OC MH AT ET NA SF TI NE OI MN		160
	TM OA FR SF HE AE IL CL EI MS UL AN NC HD AS		181
	SO RT PK NI AT FT AS EA BN LD ON DE VA ID SA		261
	ML AE LO AD NI MV RE US		298
	MK DA RL GT EA		316

a

The first letters of the six countries whose areas are shown are given, but not necessarily in the correct order.

Which country does each area represent?

c   d

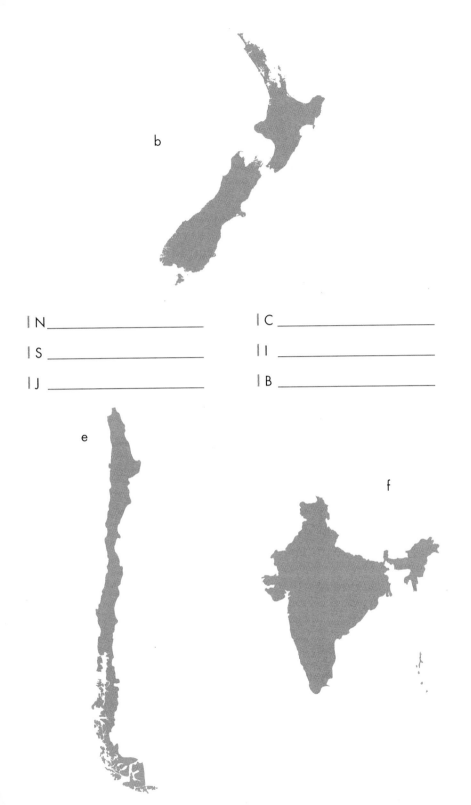

b

| N_____
| S _____
| J _____

| C _____
| I _____
| B _____

e

f

45 CANADIAN PROVINCES AND TERRITORIES

The statements below describe how the thirteen Canadian provinces and territories were named. Work out which description applies to each, but beware as one of the statements is false and will not be used.

1 Queen Victoria named this territory when it came under her jurisdiction in 1858, after a river that was in turn originally named after an explorer.

2 This province was named in honour of the fourth child of Queen Victoria.

3 This province is named after the Algonquin word for 'narrow passage' or 'strait'.

4 The name of this province originated with a Cree word used to describe part of a certain lake.

5 This province's name honours King George III, who held the title of Duke in an area with the same name in Germany.

6 The first part of this name was coined by Henry VII of England in 1497, and the second part is the name of a 15th century Portuguese explorer.

7 This name is a simple description of the territory's location.

8 This province is named after the mistress of one of the first Europeans to explore the country.

9 This province was known to the First Nations as 'Mi'kma'ki', and now has a Latin name referencing a particular region of the United Kingdom.

10 This territory's name means 'our land' in the Inuit language of Inuktitut.

11 This province was named after an Iroquois word meaning 'sparkling water'.

12 The earliest recorded name for this province is 'Abeqweit', a Mi'kmaq name meaning 'cradled in the waves'. It was renamed after a member of the British royal family in 1799.

13 A Cree name meaning 'swift-flowing river' gives this province its name

14 This territory is thought to be named after a word meaning 'great river'.

Outlines of twelve of the fifty US states are shown below, along with their nicknames. Identify each state.

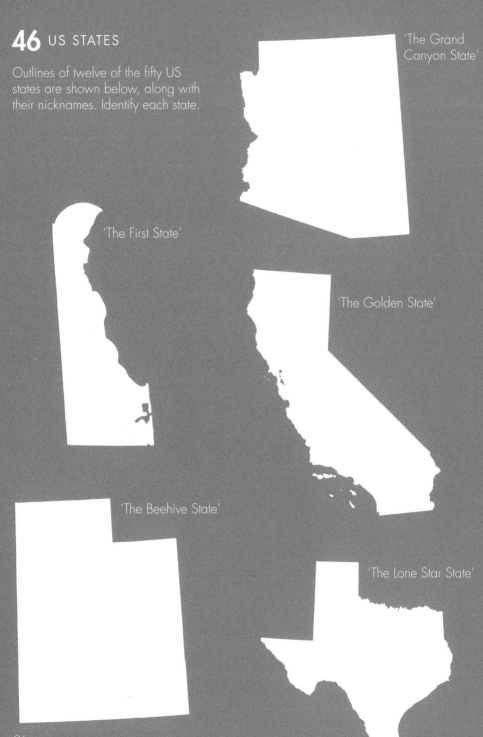

'The Grand Canyon State'

'The First State'

'The Golden State'

'The Beehive State'

'The Lone Star State'

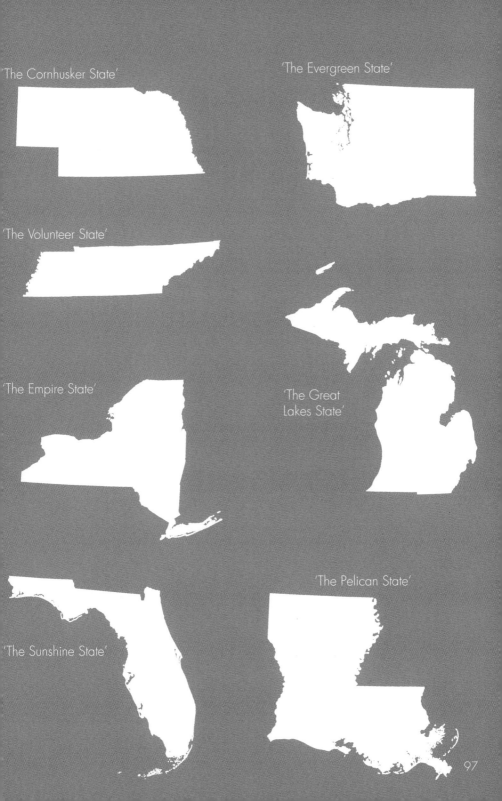

'The Cornhusker State'

'The Evergreen State'

'The Volunteer State'

'The Empire State'

'The Great
Lakes State'

'The Pelican State'

'The Sunshine State'

47 TERRITORIES QUIZ

1 Can you name all nine states and internal territories of Australia? The first letter of each state/territory, or the first letter of its first word where appropriate, is given below:

| A

| J

| N

| N

| Q

| S

| T

| V

| W

2 The Faroe Islands are an autonomous territory of which country?

3 In which Canadian province can you find Ottawa, the country's capital?

4 In which year did Hawaii become the fiftieth state of the USA?

5 Which festively named island is an Australian territory?

6 Which United States Minor Outlying Island was Amelia Earhart heading for when her plane vanished in 1937?

7 Which British Overseas Territory in the southern hemisphere has a national day dedicated to the toothfish?

8 Which uninhabited Norwegian territory is the most remote island in the world?

9 Which dependency of New Zealand shares its name with a London non-governmental organization which promotes analysis of international affairs?

10 Which territory is sometimes referred to as the '51st state'?

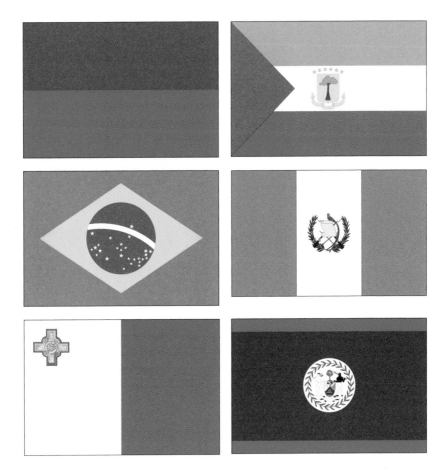

1 Find all of the national flags that feature both a crescent moon and at least one star. The first letter of each country is given, to help you out:

| A_____ | A_____
| C_____ | L_____
| M_____ | M_____
| P_____ | S_____
| T_____ | T_____
| T_____ | U_____
| C_____

2 Match the motto to the flag it is found on. The flags are shown opposite.

| Guatemala | Libertad 15 de Septiembre de 1821
| Malta | L'Union Fait La Force
| Brazil | Ordem e Progresso
| Belize | Sub Umbra Floreo
| Equatorial Guinea | Unidad, Paz, Justicia
| Haiti | For Gallantry

3 Match each bird to the country whose flag it appears on.

| Bird of Paradise | Dominica
| Grey-crowned Crane | Mexico
| Sisserou Parrot | Papua New Guinea
| Eagle | Uganda

4 Which flags do the following emblems appear on?

| A maple leaf
| A depiction of the Angkor Wat palace
| A trident

5 On which flags can you find the following unusual creatures?

| A double-headed eagle (three countries feature this)
| A dragon
| A lion holding a sword

49 CAPITAL IDENTIFICATION

On the opposite page is a map of
Europe with fifteen capital cities labelled.
However, they are all missing their
names. Match the names from the list
below to the locations on the map.

| Athens

| Bucharest

| Budapest

| Chişinău

| Kiev

| Lisbon

| Minsk

| Oslo

| Paris

| Rīga

| Sarajevo

| Vienna

| Vilnius

| Warsaw

| Zagreb

a

n

m

l

k

o

50 ODD CAPITAL OUT

Each of the following lists of capital cities all have something in common, apart from one. Can you identify the odd one out in each case?

1 | Copenhagen
  | Djibouti
  | Santo Domingo
  | Phnom Penh

2 | Reykjavík
  | Havana
  | Bogotá
  | Antananarivo

3 | Abuja
  | Khartoum
  | Kampala
  | Sofia

4 | Harare
  | Santiago
  | Port Moresby
  | Tōkyō

5 | Suva
  | Santiago
  | Port-au-Prince
  | Valletta

51 CAPITAL CITIES QUIZ

1 Each of the following is an anagram of a capital city. Can you unscramble them all? (Ignore the spaces, which are not part of the city names).

  | Gear Up

  | Mad Master

  | Mock Sloth

  | Penance Hog

  | Bran Race

  | Beta Spud

  | Lentil Gown

2 Brazil changed its capital city in 1960 as part of a plan to encourage population growth in the interior of the country. What was the capital city before the move, and where did it move to?

3 Which other capital city is nearer to Stockholm as the crow flies – Copenhagen or Helsinki?

4 In which state or territory is the capital of Australia located, and what is its name?

5 What is the state capital of New York, USA?

6 In which capital city can you find the Temple of Heaven, a UNESCO World Heritage Site?

7 Which former Russian capital was named 'Leningrad' from 1924 to 1991?

8 And which capital was known as 'Christiania' from 1624 to 1925?

b

a

The first letters of the five countries whose areas are shown are given, but not necessarily in the correct order.

Which country does each area represent?

c

d

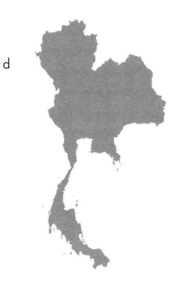

N_____	M_____
D _____	T _____
T _____	

e

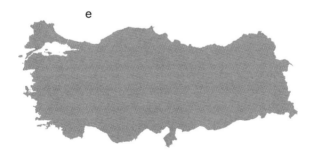

53 MONETARY MATCHING

The fifteen monetary units listed below are each used in one of the countries labelled on the map of the world opposite.

Identify each of the highlighted countries, and say which of the following units it uses.

| Dinar

| Dollar

| Euro

| Forint

| Krone

| Kuna

| Leu

| Peso

| Pound

| Real

| Rupee

| Franc

| Won

| Yen

| Yuan

a

o

54 CURRENCY QUIZ

1   In British informal usage, how much cash would you have in a 'monkey'?

2   What is sometimes referred to as a 'greenback'?

3   How many different denominations of banknote are used for the euro?

4   Which British banknote carries an image of the author of Pride and Prejudice?

5   What is the minimum number of British coins you need to make up £3.87?

6   And what is the minimum number of US coins you need to make up $3.87?

7   If you had a modern commemorative 'eagle', what value of US coin would you hold?

8   What is the highest value of coin issued by the Royal Mint in the UK?

9   What do the three-letter ISO abbreviations for Peruvian, Canadian, Colombian, Cuban, Georgian and Albanian currencies all have in common?

10 British 1p coins are only legal tender when gathered together up to a certain value – what is that limit?

55 CURRENCY ABBREVIATIONS

There are around 180 currencies recognized as legal tender around the world, and for ease of identification each is assigned a three-letter ISO abbreviation.

Ten different currency abbreviations are listed below. Match each to the full name of its associated currency.

Abbreviations	Currencies
ARS	Algerian dinar
AWG	Argentine peso
BYN	Aruban florin
CHF	Belarusian ruble
DZD	Cambodian riel
KHR	Macanese pataca
LKR	Samoan tālā
MOP	South African rand
WST	Sri Lankan rupee
ZAR	Swiss franc

OCEANIA

56 NEW ZEALAND

1 Can you find a mountain that shares its name with a city in Tuscany, famous for its leaning tower?

2 What kind of landscape is the highest mountain in New Zealand surrounded by?

3 The South Island, shown here, is famous for its beautiful fjords to the southwest. From which country, over 17,000 km from New Zealand, does the word 'fjord' originate?

4 Which two provinces of New Zealand would you travel through if you took a train from Mataura to a city named after the Gaelic name for Edinburgh?

5 Find a cutlery collection, and a place named after an equal number of rivers. If you travelled directly from one to the other as the crow flies, which water feature would you pass over at the halfway point?

6 Start at a pool named after a pale mineral, near the point where three provinces meet. From here, travel westwards to a popular New Zealand tourist destination, which Rudyard Kipling declared to be the 'eighth wonder of the world'. From here, travel south-west to what sounds like a burial place. What is the height of the mountain immediately west of your final position?

57 AUSTRALIA

1 Where on the map can you find a town that shares its name with New Zealand's capital?

2 Which four Australian states/territories are shown on this map?

3 Which town, located west of its state capital and north of the national capital, sounds like somewhere a baggage handler might visit to relax?

4 Which mountain, the highest in Blue Mountains National Park, becomes 'incorrect' when it loses a letter?

5 Which group of ancient underground caverns is located a short distance to the southeast of a place sharing its name with Shakespeare's king of the fairies?

6 From a town named after a citrus fruit, take a train in an easterly direction until you reach the first town within a built-up area marked on the map. If you follow the river that runs through this town in a northerly direction, you will pass two neighbouring towns which share their names with towns on the River Thames. What are their names?

58 PAPUA NEW GUINEA

1 Can you find an agricultural site of specific interest? It has been continuously worked for at least 7,000 years.

2 Can you find three locations on the map that each share their name with one of the following?

 | A heavily populated region of the US state of New York

 | An historical instrument for measuring the height of stars above the horizon, which was once also used for navigating

 | A type of Polynesian dance

3 Where can you find the names of four small mammals in close proximity to one another?

4 Part of Papua New Guinea was occupied by Germany in the 1880s. Which mountain range is named after the then-Chancellor of the German Empire, famous for his conquests and diplomacy?

5 Which location on the map requires only one vowel to be changed to become the name of a Middle Eastern country with one of the highest, most ancient still-populated capital cities in the world?

6 Start at a deceitful-sounding inlet and travel east along the edge of the coast until you find a Christmas key. Jump to the mainland and take the main road inland to a regional airport. East of you is a tall man sandwiched between two tall women. What is his name?

ASIA

59 PHILIPPINES

1 Which location on the map shares its name with the capital city of Chile?

2 The Philippines is made up of over 7,000 islands, separated by various seas, straits and channels. How many bodies of water are labelled as a 'sea' on the map (including partly shown names)?

3 Which volcano, north of the capital of the Philippines, became the site of an eruption with a Volcanic Explosivity Index of 6 in June 1991, making it the 20th century's second-largest eruption?

4 Which town, on the island of Tablas, is 'cool' when seen from a different direction?

5 Start at a headland on the east coast of Luzon which takes its name from the Spanish word for 'charm', but which sounds more 'enchanting' in English. Cross to the west side of the island, to a point where you might find an alligator's cousin. From here, head inland until you reach a secondary road and follow it south along the coast until you reach an airport. From here, fly due south. What is the name of the first island group you pass over?

6 Begin to the southwest at a sharp-sounding mountain named after a famous queen of Egypt. Travel along a secondary road to the northernmost tip of the island, then leap over two straits to an island with three airports. Which town on this island is one of four with the same name on the map?

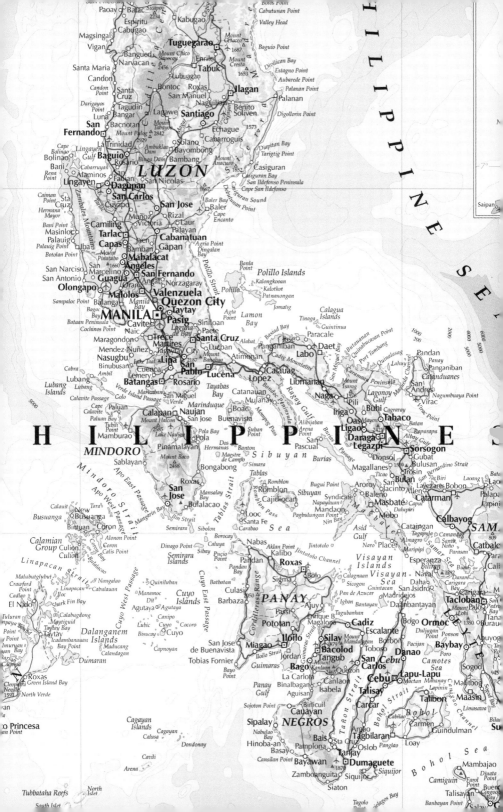

60 JAPAN

1 Which town is nearest to the point where four administrative boundaries meet?

2 What is the height of the highest non-volcanic peak marked on the map?

3 Can you find the following locations on the map that, with a single letter changed or inserted, become common words in English?

  | With an extra vowel, this place will share its name with a famous waterfall on the US-Canadian border

  | When a Z becomes an S, you will find a popular Japanese food

  | With a single letter changed, this town becomes a horseradish-based condiment often served with the solution to the above question.

4 Can you find a city that shares its name with a bright-green, melon-flavoured liqueur? It is also the Japanese word for 'green'.

5 Which town, located on the banks of the second-longest river in Japan, with which it shares its name, is also a word for a musical sound?

6 Which word, found in more than one location on the map, is the Japanese word for 'cherry blossom'?

61 CHINA

1 Which major river runs through the centre of Wuhan?

2 Hubei province is noted for its large number of lakes. Looking at the southern portion of the map, what do you think the Chinese word for 'lake' is?

3 Which river would you have to cross if you travelled in a straight line between Jiugong Shan and Wumei Shan?

4 Which location west of Wuhan shares its first three letters with the name of the dominant ethnic group in Hubei, as well as across China as a whole?

5 If you replace a 'g' with a space, which location on the map becomes a menacing group of female chickens?

6 Remove a final vowel from a location in Hubei province to find a twenty-four-hour period. From here, travel south along the motorway until it crosses another motorway, then turn right onto that other motorway. Keep going until you reach an unfinished motorway to your left. What is the name of the 'moulding' town you just passed by?

62 INDIA AND SRI LANKA

1 How many administrative provinces are there in Sri Lanka? All are shown on the map.

2 Which ancient language, also an official language of Sri Lanka and Singapore, shares its name with the first word of an Indian state labelled on the map?

3 Which location on the map shares its name with a city in Massachusetts, featured in Arthur Miller's play *The Crucible*?

4 Can you find an Indian town that has the same name as the place where Jesus spent his childhood, according to the New Testament?

5 Which map location sounds like an entrance ticket for a very large mammal?

6 Begin at a Sri Lankan headland which sounds like a disallowed goal. Follow the coast anti-clockwise until you reach a coral reef belonging to the first man. Cross the gulf by swimming atop the reef and then leave the water and join the nearest main road, heading along it in a northwesterly direction until you reach a city of 1–5 million people. Where are you now?

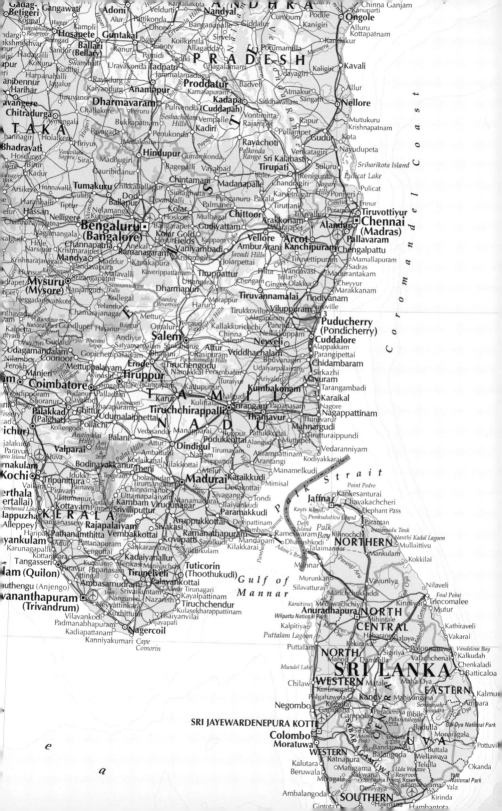

63 UAE AND OMAN

1   Which series of man-made structures in the shape of a tropical plant is labelled on the map as a site of specific interest?

2   A large proportion of the land on the Arabian Peninsula is desert, but is it composed mostly of sand or rock?

3   What do the Saudi Arabian locations of Al Quraynī, Fidā', Bir Hādī and Shahr all have in common?

4   What type of landscape is Umm as Samīm?

5   Start at Ṣuḥār and travel along the coast to Oman's northernmost enclave. From here, cross a strategically important strait to what is Iran's largest island. What is the most westerly settlement on this island?

6   Start at a site of specific interest where an antelope would be able to find respite, then travel north until you have reached the highest mountain on the map. From here, join the main road just south of the mountain range, and follow it in a northwesterly direction until you see an ancient settlement on your left, marked as a further site of specific interest. What is its name?

64 KYRGYZSTAN

1 Kyrgyzstan gained independence as a nation in 1991. Of which large Eurasian nation was it part of immediately prior to this point in history?

2 What is the difference between the water in the lakes of Ozero Balkhash (partially visible at the top-centre of the map) and Ysyk-Köl?

3 Can you find a mountain which, were it to gain an extra final letter, would share its name with a supersonic passenger aircraft?

4 In which country is the province of Badakhshān located?

5 Which riverside town in Uzbekistan, west of Kyrgyzstan, has the same name as a genre of music?

6 Find an infamous communist leader with a high geographical position on the Kyrgyzstan border. From here, head south and pass through a UNESCO World Heritage Site, home to snow leopards and Siberian ibex, until you reach what sounds like another communist peak. Where are you now?

EUROPE

65 FINLAND AND RUSSIA

1 Which city is the capital of Finland closer to: the capital of Russia, or the capital of Estonia?

2 Which town, located southeast of St Petersburg, is named after the Russian poet who wrote *Eugene Onegin*?

3 Which Finnish town, situated on the edge of a marsh, shares its name with a mounted team sport?

4 Across which two federal subjects of Russia can the largest lake in Europe be found?

5 Start at the city that was named Leningrad from 1934–1991. Travel due east by train and get off at the next town you reach. Follow the river with the same name as the city where you alighted in a southerly direction until you reach a lake. Which Russian federal subject are you in?

6 Start at a Latvian town which shares its name with an island nation located between Sicily and the African coast. Head east until you hit an international border, then follow this north through two major lakes to the coast. What is the name of the area of sea you are now standing immediately in front of?

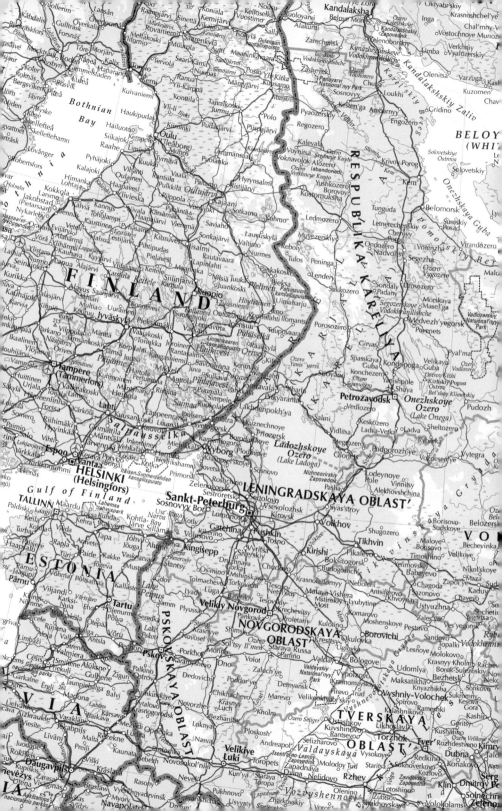

66 NORWAY

1   What is the name of the mountainous area that is home to Galdhøpiggen, Scandinavia's highest mountain at 2,469 m?

2   The Faroe Islands are shown in the map inset. Of which Scandinavian kingdom are the Faroe Islands currently a self-governing part?

3   Which valley would you traverse if you walked in a straight line from Lillehammer to Dombås, in Oppland county?

4   Can you find three locations which all start with the same letter, and fit the following descriptions? Each location is one letter longer than the previous answer.

   I A town in the north of the Sør-Trøndelag county which shares its name with a horse colouring

   I A village on the island of Vigra with the first name of a famous children's author of Norwegian heritage

   I A village in the Høgkjølen-Bakkjølen nature reserve

5   With how many countries does Norway share a land border? Not all are shown. What are those countries?

6   Start at the only regional airport in Oppland and travel east to a river that shares its name with an active Sicilian volcano. Follow the course of this river downstream until it widens into Norway's fourth-largest lake. Which town, located on the first main road east of this lake, shares the same name as the river but with just one letter changed?

67 UNITED KINGDOM

1 How many National Parks are labelled on the map?

2 Which Somerset town, shown on the map, is famous for both its Tor and its music festival?

3 Can you find locations fulfilling the following clues?

| A 'positive' peak in Devon

| A Welsh promontory which you might not be able to hear particularly clearly, were it to speak

| A place for a particular man to wash, located near the source of the river that gives Exeter its name

4 Can you identify two nearby islands, one of which is Welsh and one of which is English?

5 Start at the main airport for the capital city of Wales, and fly across the water to the nearest main airport in England. Next, make your way southwest over a county border to a town famous for its cheese and its gorge. What chops by, just to the south?

6 Start at the home of the Clifton Suspension Bridge, and travel southeast to the city formerly known as Aquae Sulis. Head east-northeast to a place that sounds good for storage, and as far south as you can to a camp by the Frome. Head due west for an ale. Where have you ended up, and what common factor links the five locations you have visited?

68 GERMANY

1 What is the name of the only labelled site of specific interest within the built-up area of Hamburg?

2 Which city on the map is pronounced in the same way as an important part of one of the methods of transport historically constructed there?

3 Which town northwest of Lübeck becomes an unwanted plant once a final letter is removed?

4 The Kiel Canal was completed in 1895 as a method of travelling between which two northern European seas without passing through the straits around the Jutland peninsula?

5 Can you find places whose names match the following descriptions and locations?

  | A clumsy person (when an umlaut is removed), near the coast

  | An item of crockery, near a large lake

  | A place that shares its name with the second-largest Norwegian city

6 From Hamburg, travel up the Elbe astride two neighbours. Head north when a third neighbour joins, leaving one behind and passing two major lakes and the founding city of the Hanseatic League. Which bay do you eventually reach?

69 THE ALPS

1   In which alpine range can you find the Matterhorn?

2   What is the name of the highest mountain in the Alps, and on the border of which two countries is it located?

3   Can you find places on the map which match the following descriptions?

   | The Italian city that hosted the 2006 Winter Olympics

   | The Swiss site of a fictional struggle between Sherlock Holmes and Moriarty

   | A word referring to a ship's prison in the valley between the Berner Alpen (Bernese Alps) and Pennine Alps

4   Places where a mountain can be crossed easily are called a 'pass' in English, or a 'col' in French. Can you find two passes that share their name with a large breed of dog, originally bred to help on alpine mountain rescues?

5   Dismount a 2,832 m-high French 'black horse', then head northeast along a river until you reach a main road. Turn left and continue along the road until you find Queen Victoria's husband within the name of a town due north of your starting point. Continue your journey to a planet circling west of your position. Where are you now?

6   Start at a town with a name similar to a festive saint, downstream from Zermatt. Follow the river north and join the motorway when you reach it, following its route west alongside a river that eventually flows into Lac Léman (Lake Geneva). Stop when you find yourself directly between an American state and a town that rhymes with the Swiss canton it is situated in. Which alpine valley is immediately south of your position?

70 GREECE

1   Which town, located towards the west of the map area, gives its name to a famous international sporting event that takes place every four years?

2   Which place beginning with 'S' was a powerful city-state in Ancient Greece, famed for its austerity and military culture?

3   Which location lends its name to an ornate order of classical architecture, as well as two books of the New Testament, and is home to the only natural land crossing from mainland Greece to the Peloponnisos (the Peloponnesian Peninsula)?

4   Can you find an archaeological site, shown as a site of specific interest, which was one of the main centres of late Bronze Age Greece, and home to the Lion Gate? It has a similar name to the nearest modern settlement to it.

5   Can you find locations on the map with names that fit the following descriptions?

    I The capital city of Libya, near the centre of the map

    I A town which is also an Italian word for 'wisdom'

    I A type of black olive, located on the coast

6   Drive along the motorway from the answer to question 3 towards Patra, stopping when you find a French bicycle that is missing its accent. From here, travel south as the crow flies, past the site of the first labour of Hercules, to a town that shares its name with a mythological many-eyed giant. What is the name of the body of water nearest to your position?

AFRICA

71 THE NILE

1 Which body of water, connecting the Red Sea and the Mediterranean Sea, was first opened as a navigable route in 1869? It takes its name from a city near the top of the map.

2 Which labels on the Nile describe two points where the river is shallow and punctuated by many boulders and rocks?

3 Can you identify the following sites of specific interest marked on the map?

| An important city in the Egyptian Old Kingdom, which shares its name with a town on the Mississippi

| A site on the banks of the Nile where King Tutankhamun's tomb is located, along with numerous other pharaonic tombs

| The original site of two temples of Rameses II carved into the side of a mountain, which were moved in the 1960s because the site was threatened by rising water

4 Which peak on the Shibh Jazīrat Sīnā' (Sinai Peninsula) shares its name with a saint after whom a medieval method of torture is named?

5 Can you find a saltwater lake that is the opposite of 'small and sweet'?

6 Start at a regional airport near the meeting point of two international borders. Dive into the body of water to the south, and swim past two protected areas to the most southerly point of a coastal national park which is a popular area for diving. From here, head southwest through an archipelago to the coast of the mainland. From the name of the protected area next to your position, what is the name of the sea that is home to this archipelago?

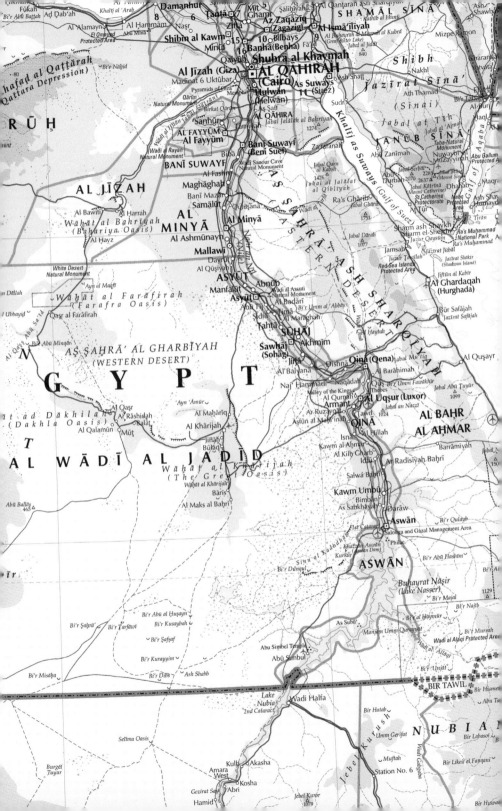

1 Based on the names of their administrative regions, which official language do you think Togo and Burkina Faso share?

2 Which African country, part of which appears on the northeast of the map, has Niamey as its capital city?

3 Which Togolese national park has a French name that translates into English as 'Lion's Pit'?

4 Which Ghanaian lake, river and administrative region all share their name with a poetic device through which a shift in tone or emotion is introduced, as well as the name of an Italian scientist who created the first battery?

5 Can you find locations on the map that match the following clues? The first letters of the answers together make an anagram of the word 'map'.

  I A Ghanaian region which shares its name with an African empire established in the seventeenth century

  I A town in Ghana that could be read as a pair of perambulators

  I A coastal region with a seafaring name

6 Start at a nature reserve that shares its western border with Côte d'Ivoire (Ivory Coast) and has a dark-sounding river running through it. Head upstream until a main road crosses your path. Follow this road in an easterly direction and take the first left, continuing until you reach a type of fish. Which animal might its name erroneously suggest you would expect to find in the national park immediately east of your position?

73 UGANDA AND TANZANIA

1 Which mountain, located in the Rwenzori range on the border of western Uganda and the Democratic Republic of the Congo (DRC), is the third highest in Africa and the highest in the DRC?

2 Can you find these two animal-related locations on the map?

  I A town on the Albert Nile in northern Uganda, which has an endangered animal in its name

  I A Tanzanian town which shares its name with a big cat that's native to the Americas

3 Which national park, famous for being the site of a large annual wildebeest migration, has a name that is derived from the Maasai word for 'endless plains'? It can be found southwest of – and across an international border from – the Masai Mara National Reserve.

4 Start at a national capital city near the banks of Lake Victoria. From here, travel east along the shoreline until you reach the point where a branch of the longest river in Africa meets the lake. Now follow the river downstream until it flows into a second lake. What is the name of this lake?

5 Begin at the city immediately west of Lake Albert. From here, head due south across an international border to find a town that sounds like a well-defended doorway. Drive south along the main road until you reach a town with a railway station. Ride the train, passing near a royal site of specific interest, then continue for some distance until you reach an airport just due west of a town that sounds like a chasm. What is its name?

6 Find a site of specific interest in Tanzania where the world's largest unfilled volcano caldera is located. From here, make your way southwest along the edge of the conservation area named after the volcano to a saltwater lake. Swim to the southwestern shore and follow the path of a river – which also forms an administrative border – southwest to a second lake. What is the name of this second lake?

1 What is the name of the most populous settlement found on the Limpopo river that is shown on this map?

2 Which province on the map gains part of its name from a South African Bantu ethnic group? The same part is one of the code words included in the NATO phonetic alphabet.

3 Which international border is closest to the South African town that shares its name with the capital of the Netherlands?

4 Which locations on the map fit the following descriptions?

  | A town on the border of a national park northwest of Pretoria (Tshwane) which sounds like it might be naturally warm and bright

  | A town near the border with Lesotho which is both a variety of flavoured tea and the name of a de facto Queen of England who ruled for nine days in 1553

  | A South African national park near the border with Lesotho which shares part of its name with a famous bridge in the city of San Francisco, California

5 Start at a tall natural landmark marked as a site of specific interest in Lesotho. Head to a different type of tall natural landmark on the international border, where it sounds like you might consume some sparkling wine. What season is contained in the name of a town just north of here?

6 Start at a South African estuary that shares its name with an island in the eastern Caribbean. Head south along the coast, passing a main airport until you have reached a coastal city with over a million inhabitants. A neighbouring settlement has a 'coniferous' name. Where are you? Can you find another town with a seemingly related name west-southwest of here?

NORTH AMERICA

75 ALASKA, USA

1 Which English city lends its name to a feature on this map?

2 Which animal, one of the largest of its type, shares its name with a national wildlife refuge and archipelago at the eastern edge of the map?

3 Can you find the following settlements on the map?

I A town, towards the north of the map, which shares part of its name with a US Founding Father

I A town, southeast of the Kuskokwim Mountains, which is also a 'goodbye'

I A palindromic town in Nushagak Bay

4 Find the mouth of the river with the same name as the Canadian territory that borders Alaska. Swim upstream past a 'town for aviators' to a 'religious burden'. On what mountain just east of your position does it sound like you might get bitten?

5 Start at the island group named after a long-tusked marine mammal. Sail north until you reach two identical natural features, then travel westwards along the coast to an inlet where you might expect to learn some pleasant information. If you sail inland along the river with the same name, which mountain do you pass on your left?

6 Begin at a river in the Aleutian Range that sounds like a regal fish. Travel upstream to the source, then cross over two volcanoes, to a 'white' peak. Head north as the crow flies to a place of suggested conflict. Where are you?

76 CANADA

1 Find locations matching each of the following descriptions.

  | A town with a palindromic girl's name, located just across a border

  | A European capital by a 'solitary lagoon'

  | A town in Minnesota which shares its name with the nocturnal creature that is the official state animal of neighbouring Wisconsin

2 If you drive directly north from the town of Carman to St Ambroise, which do you cross more of: rivers or railway lines?

3 Hop on a northbound train at Stonewall and travel to the end of the line. If you travel due east, which 'inebriated' headland do you come to?

4 There are four 'cervid' cousins in Lake Winnipeg. Can you find them all?

5 Catch a flight due north from a main airport that passes over two headlands, one of which is almost a Shakespearian tragedy and the other of which could describe some of the same play's characters. Bear right to land at the regional airport named after the river it sits on. Which island is opposite you, as you look west?

6 Take the train northwest out of Winnipeg and alight at an 'antler'. Carry on along the main road in the same direction as the train until you reach a river, then dive in and swim northeast until you reach a lake named after a French saint. Swim the length of the lake in a northeasterly direction, joining the river of the French heir, until eventually you reach a 'fishy' harbour. From here, follow the coastline northwest until you reach a named headland. It's time to show your moves – where are you?

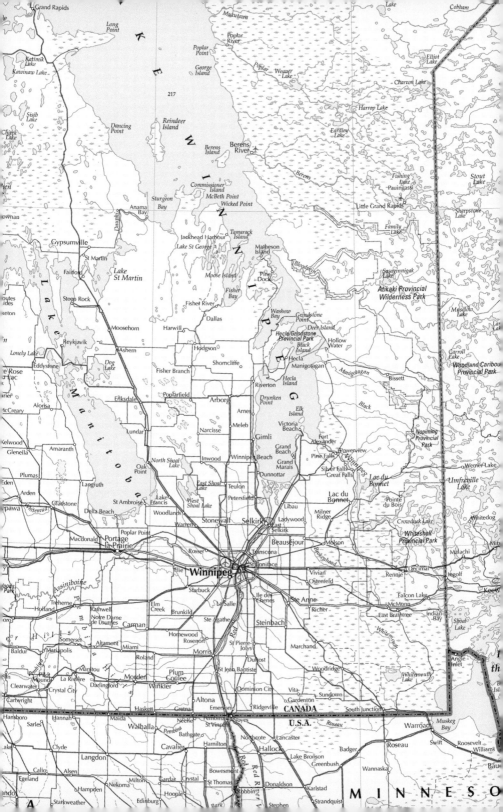

77 NEW YORK, USA

1 Which famous landmark near the mouth of the Hudson river was partly built by the same engineer who created the Eiffel Tower?

2 Which two New Jersey counties form the name of a body of water that features on the map when only the first words of each are kept?

3 Can you find the following locations, which are all on, or very close to, the same line of longitude?

   | A waterway west of the Hudson that sounds like it might be found inside an oyster

   | A seaside town that shares its name with a town on the east coast of England

   | A county that shares its name with Norway's second city

4 Start at a 'supernatural' isle at the mouth of the Mullica river. If you take the only labelled road north and turn left at every opportunity shown, which town do you end up at when you reach the end of the road?

5 Board a train in New Jersey at a place where it sounds like you might find grapes growing, and travel to a place named after a major Andalusian city. From here, take the main road southeast past the Spanish 'good', and on towards the coast. What is the name of the 'agreeable town' you reach as you first enter the marked built-up area?

6 Imagine a line connecting the capital city of Sweden to a man who served as US President during the Second World War, and who can now be found in Monmouth County. Now imagine another line, from the airport near the resting place of Viking warriors, to an Ivy League university that's home to some Tigers. Which river can be found just north of the point where these lines intersect?

1 Which location shares its name with a small citrus fruit in the mandarin family?

2 Can you find a location on the map that sounds like it would be resistant to the formation of ice crystals?

3 Can you find places on the map that match each of the following descriptions:

| The capital city of the Australian state of Victoria

| A European city famed for its Winter Palace

| A 'city of bridges', traditionally navigated using a gondola

| The third largest city in Italy

| The Roman goddess of love

4 Where's Waldo?

5 Imagine drawing a triangle that connects the three towns that could also have these definitions:

| A river flowing from the Adirondack Mountains to New York City

| A city in the Scottish Highlands, northeast of Loch Ness

| A town that, if split into two words, would mean 'gentle breeze' and 'large mounds'

What is the only five-letter town within this triangle?

6 From a black bird's deception, travel to the opposite side of Bloom until you reach a canonized mist. If you now drive towards the coast along a main road, what is the first town you reach?

1 What can you tell about the relief of the sea floor to the southwest of Guatemala from this map?

2 Can you find the following locations, which are all above the 16° line of latitude?

  | A Mexican state that lends its name to a variety of peppers, which in turn are used to make a popular brand of hot sauce

  | A Mexican coastal town whose name appears – with an extra letter – on the Brazilian flag

  | An island off the coast of Belize named after a valuable waxy substance, produced in the stomachs of whales, and used in the perfume industry

3 Start at the major city due south of the answer to question 2(b). Take the train southwest until you reach a town on the river that flows out to the Punta Frontera. If you paddle upstream instead, what role does the river take on?

4 Hop off a flight at the airport on the largest island on the map, and travel to the town on the south of the island. On the mainland west and slightly south of your position is the historical site of a Mayan walled city. What is its name?

5 Start at the national park where you might find a striped animal and head in a southeasterly direction to a town whose name translates into English as 'Liberty'. Take the road northeast and continue until you reach a coastal city that shares its name with the country in which it is found, although it is not its capital city. What direction is the country's capital city from here?

6 Start at the summit of the highest volcano on the map. Descend in a northerly direction towards the source of a river. Follow its course and cross over a national border until you reach a place to swim, which shares part of its name with a popular brand of 'bitter' drink. Swim to the northwestern end of this reserve, and rejoin the river travelling downstream until you reach a city with a population of more than half a million people. What is the height of the volcano due north of your position?

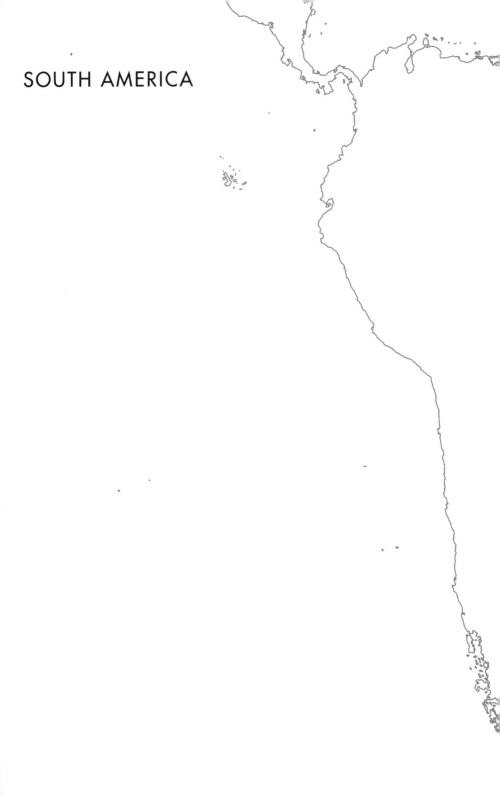

SOUTH AMERICA

80 COLOMBIA

1 Which is the nearest ocean to the city of Quibdó?

2 Which neighbouring national park, mountain range and town all share their name with a dance that was popularised in the 1990s?

3 Can you find:

 | A Colombian department which shares its name with a coastal city in northern Spain, as well as a bank?

 | A town in the Guaviare province whose name means 'the golden one', and which shares its name with the legendary city that was the subject of a poem by Edgar Allan Poe?

 | A labelled site of specific interest, in a mountain range, where a large group of megalithic sculptures can be found?

4 Imagine a line connecting the capital of Colombia with a location due west of it that shares its name with a Caucasian country. What is the name of the river that this line crosses, which lends its name to an administrative department found on the map in the north of Colombia – and which is also a Spanish girl's name?

5 In the department with a name similar to a title of Roman emperors, there is only one labelled town that sits directly on the train line. Take a railway journey from here, travelling south and taking the right-hand fork when you reach it, away from the river. When you arrive at the country's second largest city, look to the south and slightly east. What is the height of the volcano nearest to you?

6 Find the name of the second-largest South American country sitting astride a southern Colombian river. Navigate the river northwards, taking the western option at the first fork, continuing past the lake and taking the western option again at the next fork. At the third fork is an unincorporated US territory in the foothills of a mountain range. Where are you?

81 BRAZIL

1 Which island, found to the west of Rio de Janeiro, has a name that means 'big island' in English?

2 Can you find the following locations on the map? Each successive answer will contain an increasing number of the initial letter of the first answer.

  I A town that sounds similar to the original home of the Olympic Games

  I A town which, with a single vowel change, becomes a US state

  I A town southwest of São Paulo which shares its name with the capital city of Sri Lanka

3 Start at a town that is one letter away from being both 'frugal' and 'brutal'. Take the main road north until you reach a city with a population of over a million people. What is the name of this city?

4 Drive east along the main road away from Uberlândia until you reach a city whose name starts with an irrational number. From here, paddle northwards along the river until you come to a city with a name similar to the home of the Golden Gate Bridge. Where are you?

5 Start at a town in the state of São Paulo which with a single vowel change becomes the name of the European country where you can find the city of Montpellier. From here head northeast, crossing the Portuguese 'great' river to reach the state capital of California. Follow the river immediately north of your position until it flows into a small lake. What is the name of the river that flows out of the easternmost corner of this lake?

6 From the most southerly city with more than a million inhabitants on the map, travel due north to an intersection of a railway line and an administrative boundary. Make your way to a lower elevation in a northerly direction, following a river which doubles as a border until you reach the first town at the intersection of your path and a main road. If you follow this road west, what animal is contained within the name of the first town you pass on your right?

82 TIERRA DEL FUEGO

1 Can you find a body of water named after Argentina's capital city?

2 See if you can answer these questions on main roads:

| How many main roads cross the border between Argentina and Chile (remaining a main road on both sides of the border)?

| Starting at El Pluma and driving in an easterly direction using only main roads, what is the name of the first coastal town you come to?

3 In which icy national park can you find the highest volcano on the map?

4 Start at the site of specific interest whose name translates from Spanish as the 'Cave of Hands'. Which town can be found due south of here on the border between Argentina and Chile, with a name meaning 'beautiful view', and whose initials match those of the English translation?

5 Imagine a line between the only Argentinian volcano marked on the map and the town in the same country which is an anagram of 'iron mates', ignoring the space. What is the name of the river this line crosses?

6 Hop off a flight at the world's southernmost commercial airport and follow the international border north until you meet the northern coast of a fiery island. Due west of you is a saint whose name is related to the most-used calendar in the world. Who is it?

OCEANS AND POLES

83 MARIANA TRENCH

1 The name of which two-piece swimsuit can be found in the Marshall Islands?

2 Which tropical line of latitude passes through Taiwan? It is not specifically named on the map.

3 Can you find locations on the map whose names match the following descriptions?

   | A sea the same colour as a New York City taxi

   | Underwater mountains pertaining to a cartographer

   | A Scotland that is both old and new

4 Which channel lends its name to an indigenous group of Australians, distinct from the Aboriginal peoples?

5 Dive in off the south coast of Hokkaidō. Follow the trenches south, until you curve round to the west at the end of the third trench. Which profundity has thrown down the gauntlet?

6 Start at the archipelago of a king described as wise in the Old Testament. Travel due north until you reach the island to rise and shine. If you look west and slightly south, which Portuguese explorer lies buried beneath the sea?

84 NORTH POLE

1 Which attentive-sounding town is the northernmost permanent settlement in the world?

2 Which is the only administrative capital to be included on the map?

3 Which town, previously known as Thule, has a palindromic, Greenlandic name that would score 25 points in Scrabble™?

4 Can you find locations that match the following descriptions around the coast of Greenland? Your answers will be in three different languages.

  | A water feature that shares its name with the sixteenth president of the United States of America

  | An island with a Danish name that translates to 'coffee club'

  | The most populous region of France

5 How many countries are represented by territory shown on the map?

6 The area commonly known as Melville Bay has a fifteen-letter-long name in Greenlandic which means 'the great dog-sledding place', and begins and ends with the same letter. What is that name?

Sea ice extent Sep. 2012

N a n s e n   B a s i n

d-Ocean Ridge

s i n

·3910

Nordaustlandet
(Norway)

S v a l b a r d
(Norway)

Spitsbergen

57

Kvitøya

Kongsøya

Edgeøya

Sørkappøya

Hopen

·26

Barentsburg
Longyearbyen

5895

G r ø n l a n d

Greenland Sea

3899

3888

Fram Strait

Yermak
Plateau

1

Station Nord

Tobias Ø

Île de France

·596

Kap Morris Jesup

Koldeklubben Ø

P e a r y
L a n d

Kong
Frederik VIII
Land

1000

500

1500

Cape
Columbia

Lincoln
Sea

Victoria Fiord

·394

Knud Rasmussen Land

2000

Kong

Alert

Nares Strait

Hans
Island

G r e e n l a n d

N

North
Geomagnetic Pole
(2018)

E l l e s m e r e   I s l a n d

Smith Sound

(Denmark)

·1329

on Sound

Hayes
Halvø

Qaanaaq

Lauge Koch Kyst

3000

Thule Air Base
(U.S.A.)

Grise
Fiord

ones Sound

S

Qimusseriarsuaq

Nuussuaq
(Kraulshavn)

2500

N D S

N

·5346

3910

B a f f i n

on Island

·304

85 ANTARCTICA

1   Which is the highest peak in Antarctica?

2   In what year was the record low sea ice logged, according to the map?

3   Can you find the following locations on the map? All three answers consist of two words each.

   | A glacier whose name differs by a single letter from a popular condiment

   | A headland whose name is composed of two items of outerwear

   | A coast that shares its name with a fictional bear who first appeared in the *Daily Express*, back in 1920

4   Can you find the following namesakes on the map? The first two answers incorporate Roman numerals into their names, and the third is rarely found without them, outside of this map.

   | A British research base named after an astronomer, who also lends his name to a periodic comet

   | An Argentinian research base named after one of the country's 'founding fathers'

   | An area named after a monarch, between a chilly place of study and an elevated locomotive

5   Which is the only other continent that the 90° line of longitude – which bisects this map – passes through before it reaches the North Pole?

6   Start at a peninsula near the Walgreen Coast named after an animal of the family *Ursidae*. Follow the edge of the coast past the 105° line of longitude until you reach a named ice shelf, then cross over the shelf to a headland that shares its name with an animal from the family *Exocoetidae*. What is the name of the island you have reached?

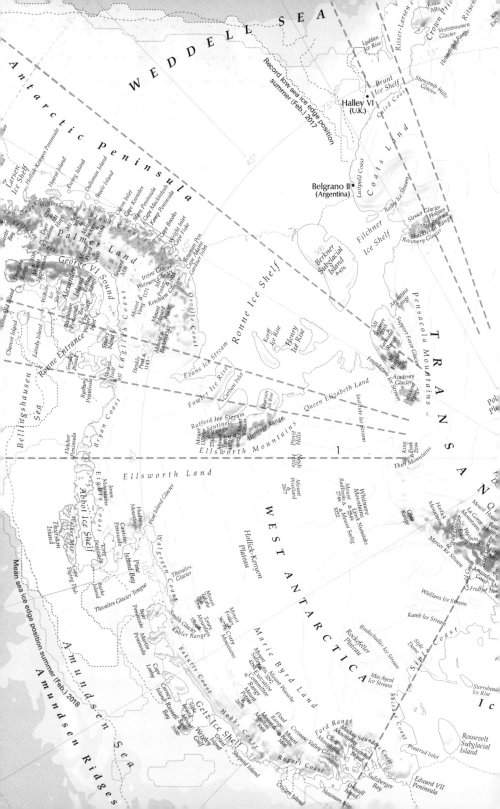

WEDDELL SEA

Antarctic Peninsula

Record low sea ice edge position summer (Feb) 2017

Lyddan Ice Rise

Riiser-Larsen Ice Shelf

Crown Pri

Brunt Ice Shelf

Stancomb-Wills Glacier

Halley VI (U.K.)

Third Coast

Coats Land

Larsen Ice Shelf
Larsen Coast

Wilkins Coast
Biscoe
Hollick-Kenyon Peninsula

Hearst Island
Steele Island
Beatrix Island
Dolleman Island

Kemp Peninsula
Cape Mackintosh
Odom Inlet

Marz Peninsula
Cape Knowles

Cape Brooks

Wright Inlet
Cape Fiske

Belgrano II (Argentina)

Luitpold Coast

Filchner Ice Shelf

Slessor Glacier
Holmes
Shackleton Range
Recovery Glacier

Palmer Land

Black Coast

Mount Hope

George VI Sound

Irvine Glacier
Wetmore Glacier
Ketchum Glacier

Orville Coast

Ronne Ice Shelf

Berkner Subglacial Island

Argentina Range

Pensacola Mountains

TRANS

Alexander Island

English Coast

Wilkins Ice Shelf

Mount Baxter

Evans Ice Stream

Korff Ice Rise
Henry Ice Rise

Support Force Glacier

Academy Glacier

Bellingshausen Sea

Ronne Entrance

Fowler Ice Rise

Carlson Inlet

Foundation Ice Stream

Bryan Coast
Fletcher Peninsula

Rutford Ice Stream
Sentinel Range

Skytrain Ice Rise

Queen Elizabeth Land

Institute Ice Stream

Ellsworth Mountains

Heritage Range

Pirrit Hills

Nash Hills

King Peak
Thiel Mountains

Mean sea ice edge position summer (Feb) 2018

Ellsworth Land

Abbot Ice Shelf

Thurston Island

Eights Coast

Hudson Mountains

Canisteo Peninsula

Walgreen Coast

Hollick-Kenyon Plateau

Mount Woollard

Whitmore Mountains Nunataks
Radford Ice
Mount Seelig

Horlick Mountains

Olga Range

WEST ANTARCTICA

Marie Byrd Land

SAN

Amundsen Sea

Pine Island Bay

Thwaites Glacier

Thwaites Glacier Tongue

Burke Island

Bear Peninsula
Martin Peninsula

Kohler Range

Smith Glacier

Crary Mountains

Mount Frakes

Executive Committee Range
Mount Waesche
Mount Sidley

Amundsen

Rockefeller Plateau

Siple Coast

Bindschadler Ice Stream

MacAyeal Ice Stream

Whillans Ice Stream

Kamb Ice Stream

Side Dome

Steershead Ice Rise

Bakutis Coast
Hobbs Coast

Getz Ice Shelf

Ford Range

Flood Range
Ames Range

Crevasse Valley Glacier

Ruppert Coast

Sulzberger Bay

Roosevelt Subglacial Island

Edward VII Peninsula

Prestrud Inlet

Amundsen Ridges

Wrigley Gulf

SOLUTIONS

PRETORIA

ESPANA

CORK

ASTI

THE WORLD TODAY

1 WEATHER
1. There were 11 major tropical storms in 2005 in the following places: Louisiana, S. Mexico, Central America, Florida/Alabama, Bahamas, Caribbean, Madagascar, Kyūshū, Taiwan, N. Coast Australia and N.W. Coast Australia.
2. Rainforest
3. They all form over water in warm regions.
4. Typhoon: Northwest Pacific region, Hurricane: North Atlantic and East Pacific regions, Cyclone: Indian Ocean region.
5. Finland: Subarctic, Greece: Mediterranean, New Zealand: Temperate, Oman: Desert, Uruguay: Humid subtropical, Indonesia: Rainforest.

2 CLIMATE GRAPHS
1. Vladivostok
2. Yakutsk
3. Kābul – 1,799m above sea level.
4. Yakutsk
5. ~45 °C. The highest average temperature in July is ~23 °C, while in December the lowest average temperature is ~-22 °C.

3 CHANGE IN PRECIPITATION
1. Oceania
2. Morocco
3. Antarctica
4. Africa, with a 50% drop in precipitation on the northwest coast, and a 50% increase in central and far-eastern Africa.

4 WEATHER EXTREMES

Record	Location	Record
Highest shade temperature	Furnace Creek, Death Valley, California, USA	56.7 °C
Hottest place (annual mean)	Dalol, Ethiopia	34.4 °C
Driest place (annual mean)	Atacama Desert, Chile	0.1 mm
Most sunshine (annual mean)	Yuma, Arizona, USA	over 4,000 hours
Least sunshine	South Pole	nil for 182 days per year
Coldest place (annual mean)	Plateau Station, Antarctica	-56.6 °C
Wettest place (annual mean)	Meghalaya, India	11.873 m

Record	Location	Record
Most rainy days	Mount Waialeale, Hawaii	up to 350 per year
Greatest snowfall	Mount Rainier, Washington, USA	31.102 m
Thunder-days average	Tororo, Uganda	251 days per year

5 WORLD POPULATION DISTRIBUTION
1. The area to the northeast of India, including Bangladesh – which is in fact the world's most densely populated country.
2. Greenland – it is largely uninhabited.
3. Oceania
4. Java – the city of Jakarta is home to over 10 million people.
5. 1, Asia; 2, Africa; 3, Europe (including Russia); 4, South and Central America, and the Caribbean; 5, Northern America; 6, Oceania

6 POPULATION GROWTH
1. 1990s
2. 1,000 million
3. Latin America and the Caribbean is projected to overtake Europe
4. Approximately 3,800 million in Asia and 500 million in Latin America and the Caribbean, giving a difference of 3,300 million.

7 POPULATION QUIZ
1. 2015
2. Europe
3. New York
4. Tōkyō
5. 2007

8 POPULATION CHANGE
1. Decreasing, as indicated by the light green colour.
2. India (14,829,000), China (5,504,000), Nigeria (4,994,000), Pakistan (3,796,000), Indonesia (2,812,000), Democratic Republic of the Congo (2,662,000), Ethiopia (2,577,000), USA (2,300,000), Egypt (1,833,000), Tanzania (1,779,000)
3. Portugal, Ukraine, Romania, Latvia, Bulgaria. The rate of change is <-0.30 % – the countries are coloured dark green on the map.

9 CITY POPULATIONS
Oceania

4,967,733	Melbourne	Australia
4,925,987	Sydney	Australia

2,406,182	Brisbane	Australia
2,041,959	Perth	Australia
1,606,564	Auckland	New Zealand

Europe

15,190,336	İstanbul	Turkey
12,537,954	Moscow	Russia
11,017,230	Paris	France
9,304,016	London	United Kingdom
6,617,513	Madrid	Spain

Asia

37,393,129	Tōkyō	Japan
30,290,936	Delhi	India
27,058,479	Shanghai	China
21,005,860	Dhaka (Dacca)	Bangladesh
20,462,610	Beijing	China

South America

22,043,028	São Paulo	Brazil
15,153,729	Buenos Aires	Argentina
13,458,075	Rio de Janeiro	Brazil
10,978,360	Bogotá	Colombia
10,719,188	Lima	Peru

Africa

20,900,604	Cairo	Egypt
14,368,332	Lagos	Nigeria
14,342,439	Kinshasa	Democratic Republic of the Congo
8,329,798	Luanda	Angola
6,701,650	Dar es Salaam	Tanzania

North America

21,782,378	Mexico City	Mexico
18,803,552	New York	United States of America
12,446,597	Los Angeles	United States of America
8,865,009	Chicago	United States of America
6,370,704	Houston	United States of America

10 MOST POPULOUS CITIES

1. 5 – Bogotá, Lima, Buenos Aires, São Paulo, Rio de Janeiro
2. Asia
3. 6 – Tōkyō, Delhi, Beijing, Mexico City, Cairo, Dhaka (Dacca)
4. Canada
5. Any two of the following: Namibia, Botswana, Lesotho, Eswatini (Swaziland).

11 CITY CLOSE-UPS

1. İstanbul
2. 4 – Delhi, Manila, Jakarta, Karachi (shortened to Nat. Mus.); Shanghai has a 'Shanghai Museum' and Manila has a 'Casa Manila Museum', as well as the 'Museum of the City of Manila'; 2 – Shanghai, Delhi
3. 2 – İstanbul and Manila – they both have two.
4. Religions indicated by places of worship, and the cities they can be found in, are as follows. One building of each religion per city is given as an example, although more may be present in some cases:
   | Islam (4 cities) – Istiqlal Mosque, Jakarta; Jama Masjid Mosque, Delhi; Memon Mosque, Karachi; Nusretiye Mosque, İstanbul
   | Buddhism (3 cities) – Chinese Temple, Jakarta; Jade Buddha Temple, Shanghai; Ladakh Buddha Vihar, Delhi
   | Hinduism (1 city) – Lakshmi Narayan Temple, Delhi
   | Jainism (1 city) – Digamber Jain Temple, Delhi
   | Christianity (various denominations) (6 cities) – San Miguel Catholic School & Church, Manila; Catholic Cathedral, Jakarta; Christ the King Church, Karachi; Hagia Sophia, İstanbul (later a mosque); Muen Church, Shanghai; St James Church, Delhi
   Counting Christianity as a whole, it is the most represented religion, with places of worship appearing on all six maps.
5. 'Great Laundry' in Karachi.

12 LIFE EXPECTANCY

1. USA and China, with life expectancies of 70–79. Over 40 other countries have life expectancies of 80+.
2. Mongolia
3. Chile has a life expectancy of 80+, while Guyana and Bolivia have a life expectancy of 60–69.
4. 60–69
5. Hong Kong and Japan.

13 DEATHS FROM POOR HEALTHCARE

1. Democratic Republic of the Congo, which shares its poor survival rate with neighbouring Central African Republic, South Sudan, Angola and Zambia.
2. Yemen and Iraq.
3. Libya, with 0–14.9% of deaths, contrasts strongly with neighbouring Niger and Chad, each of which have rates of 60% and above.
4. Mongolia and Turkmenistan.
5. Afghanistan, with a death rate of 39% of the population. The next worst country outside Africa is Laos, with a death rate of 35% of the population.

14 HEALTH QUIZ

1. It increased from 30 years to 70 years.
2. Around a third.
3. 8 times
4. Disability Adjusted Life Year.
5. 1980
6. 60%
7. Argentina and Algeria.
8. Finland
9. Vaccination
10. Antibiotic

15 TRADE

1. Lesotho
2. 8 – Morocco, Tunisia, Somalia, Malawi, Botswana, Equatorial Guinea, Ghana, Sierra Leone
3. Canada, the only country from the list above 50%.
4. Oceania
5. Luxembourg, at 407% of its GDP.

16 INCOME INEQUALITY

1. Brazil
2. North America and Africa.
3. Japan, Afghanistan and Azerbaijan.
4. Europe
5. Namibia, Honduras, Suriname, Lesotho and Eswatini (Swaziland).

17 GROSS NATIONAL INCOME

1. Singapore and Brunei,
2. The Middle East – Qatar, Kuwait and the United Arab Emirates.
3. USA
4. Norway – it is in the lowest income inequality, while having a GNI per capita in excess of $60,000.
5. Luxembourg. The other two European countries in this band are Switzerland and Norway.

18 MOBILE PHONE USERS

1. Russia
2. Botswana, Gabon and Libya.
3. The USA has 355,500,000 cellular subscribers, and a density of 90 to 119.9 phones per 100 inhabitants. Therefore, its population is between 319.95 million and 426.24 million people.

4. Any two of the following: Kosovo, Europe; Western Sahara, Africa; and French Guiana, South America.
5. South Sudan, Africa

19 TELECOMMUNICATIONS
1. 2006
2. Around 100%, from just under 2,000 million at the start of 2004 to 4,000 million in 2008.
3. At the start of 2007.
4. 11,470,000 since Africa accounts for 1% of the world's main telephone lines, of which there are 1.147 billion in 2014.
5. Broadband subscribers in the Americas, since there were 23.1% x 3.026 million = 699,000 Americas broadband subscribers, versus 1% x 6.915 million = 630,000 African mobile cellular subscribers.

20 COMMUNICATIONS QUIZ
1. a. Longest railway tunnel   b. Longest road tunnel
   c. Longest railway   d. Longest wall
   e. Longest suspension bridge   f. Longest bridge
   g. Tallest dam   h. Tallest building
   i. Longest road   j. Tallest road bridge
2. 1956. The cable was known as 'TAT1'.
3. Antarctica
4. 200,000
5. 36%, although this represents a drop from 2001, when North America hosted 69% of the world's servers.
6. Europe
7. Less than 1% (in particular, 0.7% as of 2018).

21 WORLD TIME ZONES
1. 09:30, since the time zone in Mexico City is UTC-6.
2. China, which consists of a single, large time zone, introduced under Mao Zedong in 1949.
3. The local time on landing would be 5pm on Saturday, given that Helsinki is UTC+2 and Port Moresby is UTC+10.
4. Russia, with 11.
5. Helsinki, Finland   Johannesburg, South Africa
    Beijing, China   Perth, Australia
    Moscow, Russia   Addis Ababa, Ethiopia
    Paris, France   Lagos, Nigeria
    Ottawa, Canada   Lima, Peru
    Reykjavík, Iceland   Dakar, Senegal

22 SUMMER AND WINTER TIMES

1. In April, Paraguay is UTC-4 and Phoenix is UTC-6, with a time difference of 2 hours. In December, Paraguay is UTC-3 and Phoenix is UTC-7, with a time difference of 4 hours. So the time difference is smaller in April than in December.
2. ACDT is UTC+10:30 and IST is UTC+2, so IST is 8:30 behind ACDT. Therefore when it is 6:45 p.m. in ACDT, it is 10:15 a.m. on the same day in IST.
3. In July, Phoenix is UTC-6 and Adelaide is UTC+9:30, so Adelaide is 15:30 ahead. Therefore when it is 12 p.m. in Phoenix, it is 3:30 a.m. the next day in Adelaide.
4. In September, Israel is UTC+3 and Paraguay is UTC-4, so Paraguay is 7:00 behind. Therefore when it is 3:30 a.m. in Israel it is 8:30 p.m. the day before in Paraguay.
5. In December, Fiji is UTC+13 and Israel is UTC+2, so Israel is 11:00 behind Fiji. Therefore when it is midnight Sunday in Fiji, it is 1 p.m. on Saturday in Israel.

23 TIME ZONE ABBREVIATIONS

ADT	Atlantic Daylight Time (USA) or Arabia Daylight Time
AEST	Australian Eastern Standard Time
CET	Central European Time
CST	China Standard Time or Central Standard Time (US)
CXT	Christmas Island Time
EAT	East Africa Time
ECT	Ecuador Time
EET	Eastern European Time
HKT	Hong Kong Time
IDT	Israel Daylight Time
JST	Japan Standard Time
MSK	Moscow Time
NT	Newfoundland Time (Canada)
NZDT	New Zealand Daylight Time
SST	Singapore Standard Time or Samoa Standard Time
WAT	West Africa Time

24 THE UNITED STATES

1. Eastern, Central, Mountain and Pacific Time.
2. Atlantic, Alaska, Hawaii-Aleutian, Samoa and Chamorro Time.
3. Atlantic, Samoa and Chamorro.
4. UTC-3

GEOGRAPHICAL INFORMATION

25 WORLD'S HIGHEST MOUNTAINS

Letters in a. to b. have been shifted forwards by one letter in the alphabet, and in c. to f. have been shifted by three letters. In the numbered codes, the letters are all assigned numbers depending on their position in the alphabet. For questions g. and h., A = 1, B = 2, and so on, while in questions i. and j., A = 26, B = 25 and so on.

a.   Mount Everest
b.   Chogori Feng (also known as K2)
c.   Kangchenjunga
d.   Lhotse
e.   Makalu
f.   Cho Oyu
g.   Dhaulagiri I
h.   Manaslu
i.   Nanga Parbat
j.   Annapurna I

26 MOUNTAIN QUIZ

1. 4 – Mt Everest, Lhotse, Makalu and Cho Oyu.
2. True
3. Edmund Hillary and Tenzing Norgay.
4. 5 – Nepal, India, Bhutan, China, Pakistan.
5. The Hawaiian volcano Mauna Kea, which is 10,000 m high when measured from its base on the sea floor.
6. Ben Nevis, Snowdon and Scafell Pike – collectively known as the Three Peaks.
7. The Andes
8. The Appalachian Mountains
9. 8,000 m – the altitude above this height is commonly referred to as the 'death zone'.
10. The border between Europe and Asia.

27 COUNTRIES AND THEIR FLAGS: EASY

1.

Spain   Greece   Barbados

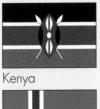

Kenya

Mexico

New Zealand

Norway

Slovakia

South Africa

2. The Sun
3. 5
4.

Switzerland

Vatican City

Nepal

5. They both prominently feature Arabic script in the centre of the flag.
6. Bahrain, Qatar
7. Poland
8.

Sweden

Finland

Norway

Denmark

28 ISLAND MATCHING

a. Madagascar
b. Borneo
c. Baffin Island
d. Victoria Island
e. Greenland
f. Great Britain
g. New Guinea
h. Ellesmere Island
i. Sumatra (Sumatera)
j. Honshū

29 ISLAND QUIZ

1. Denmark
2. 3 – Victoria Island, Ellesmere Island, Baffin Island.
3 Borneo – Malaysia, Indonesia and Brunei share the island (Scotland,

Wales and England share Great Britain, but are not considered to be independent of one another).
4. Papua New Guinea and Indonesia.
5. The second most northerly. Honshū is located north of both Kyūshū and Shikoku but south of Hokkaidō.

30 WORLD'S LARGEST ISLANDS

Island	Area
Greenland	2,175,600 sq km
New Guinea	808,510 sq km
Borneo	745,561 sq km
Madagascar	587,040 sq km
Baffin Island	507,451 sq km
Sumatra (Sumatera)	473,606 sq km
Honshū	227,414 sq km
Great Britain	218,476 sq km
Victoria Island	217,291 sq km
Ellesmere Island	196,236 sq km

31 WORLD'S LARGEST LAKES

Lake	Bordering countries
Caspian Sea	Kazakhstan, Turkmenistan, Iran, Azerbaijan, Russia
Great Bear Lake	Canada
Great Slave Lake	Canada
Lake Huron	Canada, USA
Lake Michigan	USA
Lake Nyasa (Lake Malawi)	Tanzania, Mozambique, Malawi
Lake Superior	Canada, USA
Lake Tanganyika	Burundi, Tanzania, Zambia, Democratic Republic of the Congo
Lake Victoria	Uganda, Kenya, Tanzania
Lake Baikal (Ozero Baykal)	Russia

32 LAKES QUIZ

1. Lake Victoria
2. 5 – Lake Superior, Lake Michigan, Lake Huron, Lake Erie, Lake Ontario.
3. 1 – the Caspian Sea.
4. Russia – Lake Baikal is approximately 1,700 m deep.
5. Russia – also Lake Baikal, which was formed around 25 million years ago.
6. Straits of Mackinac

33 LAKE AREAS

Lake	Area
Caspian Sea	371,000
Lake Superior	82,100
Lake Victoria	68,870
Lake Huron	59,600
Lake Michigan	57,800
Lake Tanganyika	32,600
Great Bear Lake	31,328
Lake Baikal (Ozero Baykal)	30,500
Lake Nyasa (Lake Malawi)	29,500
Great Slave Lake	28,568

34 IDENTIFY THE COUNTRY: EASY

a. Italy b. China c. Australia d. United States of America

35 WORLD'S LONGEST RIVERS

a. Mississippi-Missouri
b. Ob'
c. Yenisey-Angara-Selenga
d. Irtysh (Yertis)
e. Yellow river (Huang He)
f. Yangtze (Chang Jiang)
g. Nile
h. Congo
i. Río de la Plata-Paraná
j. Amazon (Amazonas)

36 RIVERS QUIZ

1. The Danube – it flows through ten countries (Germany, Austria, Slovakia, Hungary, Croatia, Serbia, Bulgaria, Romania, Moldova, Ukraine) before draining into the Black Sea.
2. The Volga – at 3,688 km long.
3. The Nile
4. The Yangtze – the Three Gorges Dam is one of the largest hydroelectric dams in the world.
5. The Colorado river

37 THE MOUTH OF THE RIVER

River	Sea it drains into
Amazon (Amazonas)	Central Atlantic Ocean, near the equator
Congo	South Atlantic Ocean
Mississippi-Missouri	Gulf of Mexico
Nile	Mediterranean Sea
Ob'	Kara Sea
Río de la Plata-Paraná	South Atlantic Ocean
Yangtze (Chang Jiang)	East China Sea

Yellow river (Huang He)   Bo Hai (a bay of the Yellow Sea)
Yenisey-Angara-Selenga   Kara Sea
Irtysh (Yertis)   Kara Sea

38 COUNTRIES AND THEIR FLAGS: MEDIUM

1.

Belarus

Panama

Seychelles

2. Bangladesh. All four are a flat colour with a symbol in the centre, but the Bangladeshi flag has a filled circle in its centre whereas the other three have stars.
3. The pairs and their colours are:

North Macedonia	China	Red and yellow
Belgium	Germany	Red, yellow and black
Ireland	Côte d'Ivoire	Orange, white and green
Hungary	Italy	Red, white and green
Russia	France	Blue, white and red
Canada	Poland	Red and white
Chad	Colombia	Red, blue and yellow
Federated States of Micronesia	Somalia	Blue and white

4. Australia, Brazil, New Zealand, Papua New Guinea and Samoa.
5. 3 – Antigua and Barbuda, Kiribati, Malawi.
6. Cyprus and Kosovo.
7. Mozambique

39 AFRICA AND EMPIRE

Name (1898)	Name (2019)
Abyssinia	Ethiopia (part of)
British East Africa	Kenya and Uganda
Cape Colony	South Africa (southernmost part)
Congo State	Democratic Republic of the Congo
French Guinea	Guinea
French Somaliland	Djibouti
German East Africa	Burundi, Rwanda and Tanzania
Kamerun	Cameroon
Portuguese East Africa	Mozambique
Rio de Oro	Western Sahara (western part of)

40 COUNTRY NAMES QUIZ
1. Union of Soviet Socialist Republics, known in Russian as CCCP.
2. Madras
3. Democratic Republic of the Congo
4 Ireland
5. Iraq

41 COUNTRY NAMES PUZZLES
1. Chad, Cuba, Fiji, Iran, Iraq, Laos, Mali, Niue, Oman, Peru, Togo
2. United Arab Emirates
3. W and X
4. Madagascar, Malawi, Malaysia, Maldives, Mali, Malta, Marshall Islands, Mauritania, Mauritius, Mexico, Federated States of Micronesia, Moldova, Monaco, Mongolia, Montenegro, Morocco, Mozambique, Myanmar
5. The United Kingdom of Great Britain and Northern Ireland

42 LARGEST COUNTRIES

Map letter	Country	Area (sq km)
c.	Russia	17,075,400
b.	Canada	9,984,670
a.	USA	9,826,635
d.	China	9,606,802
j.	Brazil	8,514,879
f.	Australia	7,692,024
e.	India	3,166,620
i.	Argentina	2,766,889
g.	Kazakhstan	2,717,300
h.	Algeria	2,381,741

43 SMALLEST COUNTRIES

Map letter	Country	Area (sq km)
c.	Vatican City	0.5
a.	Monaco	2
g.	Nauru	21
i.	Tuvalu	25
h.	San Marino	61
b.	Liechtenstein	160
e.	Marshall Islands	181
j	St Kitts and Nevis	261
d.	Maldives	298
f.	Malta	316

44 IDENTIFY THE COUNTRY: MEDIUM
a. Brazil b. New Zealad c. Spain d. Japan e. Chile f. India

45 CANADIAN PROVINCES AND TERRITORIES
1. d. British Columbia
2. e. Alberta
3. i. Quebec
4. g. Manitoba
5. j. New Brunswick
6. m. Newfoundland and Labrador
7. b. Northwest Territories
8. false
9. l. Nova Scotia
10. a. Nunavut
11. h. Ontario
12. k. Prince Edward Island
13. f. Saskatchewan
14. c. Yukon

46 US STATES
The First State: Delaware; The Beehive State: Utah; The Grand Canyon State: Arizona; The Golden State: California; The Lone Star State: Texas; The Cornhusker State: Nebraska; The Volunteer State: Tennessee; The Empire State: New York; The Sunshine State: Florida; The Evergreen State: Washington; The Great Lakes State: Michigan; The Pelican State: Louisiana

47 TERRITORIES QUIZ
1. Australian Capital Territory, Jervis Bay Territory, New South Wales, Northern Territory, Queensland, South Australia, Tasmania, Victoria, Western Australia
2. Denmark
3. Ontario
4. 1959
5. Christmas Island
6. Howland Island
7. South Georgia and South Sandwich Islands
8. Bouvet Island – which is almost 1,700 km away from the nearest land.
9. Chatham Islands. Chatham House is the name of the London-based NGO.
10. Puerto Rico

48 COUNTRIES AND THEIR FLAGS: TRICKY
1. Algeria, Azerbaijan, Comoros, Croatia, Libya, Malaysia, Mauritania, Pakistan, Singapore, Tunisia, Turkey, Turkmenistan and Uzbekistan.
2. Guatemala   'Libertad 15 de Septiembre de 1821'
    Haiti   'L'Union Fait La Force'
    Brazil   'Ordem e Progresso'

Belize	'Sub Umbra Floreo'
Equatorial Guinea	'Unidad, Paz, Justicia'
Malta	'For Gallantry'

3. Bird of Paradise   Papua New Guinea
    Grey-crowned Crane   Uganda
    Sisserou Parrot   Dominica
    Eagle   Mexico

4. Canada; Cambodia; Barbados
5. Albania, Serbia and Montenegro; Bhutan; Sri Lanka

49 CAPITAL IDENTIFICATION

a. Oslo b. Riga c. Vilnius d. Minsk e. Kiev f. Chişinău g. Budapest
h. Bucharest i. Sarajevo j. Athens k. Zagreb l. Paris m. Vienna
n. Warsaw o. Lisbon

50 ODD CAPITAL OUT

1. Phnom Penh – all of the others are capital cities of a country beginning with 'D'.
2. Bogotá – the others are all capitals of island countries.
3. Sofia – the others are all capital cities of African countries.
4. Tōkyō – all of the other capital cities are in the southern hemisphere.
5. Suva – all the others are capitals of five-letter countries.

51 CAPITAL CITIES QUIZ

1. Prague, Amsterdam, Stockholm, Copenhagen, Canberra, Budapest, Wellington
2. Rio de Janeiro was the capital until 1960, when it changed to Brasília.
3. Helsinki
4. Australian Capital Territory, and its name is Canberra.
5. Albany
6. Beijing, China
7. St Petersburg
8. Oslo

52 IDENTIFY THE COUNTRY: TRICKY

a. Mexico b. Denmark c. Norway d. Thailand e. Turkey

53 MONETARY MATCHING

a. Real: Brazil; b. Kuna: Croatia; c. Krone: Norway; d. Forint: Hungary; e. Leu: Moldova; f. Yen: Japan; g. Won: South Korea; h. Dollar: Australia; i. Yuan: China; j. Rupee: India; k. Euro: Cyprus; l. Pound: Egypt; m. Franc: Rwanda; n. Dinar: Algeria; o. Peso: Argentina

54 CURRENCY QUIZ

1. £500

2. A US banknote, often specifically a dollar bill. Originally, 'greenbacks' were paper currency issued by the US during the American Civil War.

3. 7 – €500, €200, €100, €50, €20, €10 and €5 (although the last €500 banknote was issued in April 2019, it remains legal tender).

4. £10 – Jane Austen features on the design of the note that was first issued in September 2017.

5. 7 coins – 1 x £2, 1 x £1, 1 x 50p, 1 x 20p, 1 x 10p, 1 x 5p, 1 x 2p

6. 8 coins – 3 x $1, 1 x 50¢, 1 x 25¢, 1 x 10¢, 2 x 1¢

7. $10 – the name 'eagle' was first assigned to this coin in the US Coinage Act of 1792.

8. £100 – although they are officially legal tender, they are classed as 'non-circulating' and generally cannot be used except for payment of court debts.

9. They all form three-letter English words: PEN, CAD, COP, CUP, GEL, ALL

10. 20p – for totals above that, they can be legally refused as payment.

55 CURRENCY ABBREVIATIONS

ARS	Argentine peso
MOP	Macanese pataca
BYN	Belarusian ruble
KHR	Cambodian riel
ZAR	South African rand
AWG	Aruban florin
CHF	Swiss franc
DZD	Algerian dinar
LKR	Sri Lankan rupee
WST	Samoan tālā

OCEANIA

56 NEW ZEALAND

1. Mount Pisa, in Otago province. Pisa is located on the northwest coast of Italy.
2. Glaciers. Aoraki/Mount Cook, the highest mountain both on the map and in New Zealand at 3,724 m, is located on the border of the West Coast and Canterbury provinces and is surrounded by glaciers, indicated by the dotted blue lines on a white background.
3. Norway. A fjord is a long, narrow and deep inlet of the sea, sandwiched between high cliffs, and usually the remains of a glacial valley.
4. Southland and Otago, since you would travel to Dunedin. Its name is derived from Dùn Èideann, Edinburgh's Gaelic name. The train line is indicated by a solid grey line, running parallel to the east coast of the island.
5. Poolburn Reservoir. The locations are the towns of Five Forks in Otago and Five Rivers in Southland.
6. 966 m (Bare Cone). Travel from Lake Alabaster (a pale mineral) to Milford Sound (actually a fjord, despite its name) to Lake Grave (which sounds like a burial place).

57 AUSTRALIA

1. Wellington, on the railway line near the top centre of the map.
2. New South Wales, Victoria, Australian Capital Territory and Jervis Bay Territory.
3. Porters Retreat, located west of Sydney and north of Canberra.
4. Mount Werong, at 1,214 m. If you take away the 'e', the second word becomes 'wrong'.
5. Jenolan Caves, just southeast of Oberon, near the centre of the map to the west of Sydney. The caves are thought to be around 340 million years old.
6. Richmond and Windsor. Travel from Orange to Penrith along the train line, then head north along the Nepean river.

58 PAPUA NEW GUINEA

1. Kuk Early Agricultural Site, just above the second 'P' of the PAPUA NEW GUINEA country name.
2. ❙ Long Island, just above the 'E' of the PAPUA NEW GUINEA country name
   ❙ Astrolabe Bay, south of Karkar Island.
   ❙ Hula, located southeast along the coast from Port Moresby.
3. The Mwilitau Islands – Bat Island, Rat Island, Mole Island and Mouse Island – located near the top of the map.
4. Bismarck Range. Otto von Bismarck was the first Chancellor of the German Empire, holding the position from 1871–1890.
5. Yamen – this is one letter away from Yemen, whose capital city is Ṣanʿāʾ (Sanaa). It can be found south of the coastal town of Darapop near the top-left of the map.
6. Mount Albert Edward. The journey starts at Deception Bay on the south coast, and heads to Yule (Christmas) Island (key), before moving inland to the mountain that is located between Mt St Mary and Mount Victoria (two tall women).

ASIA

59 PHILIPPINES

1. Santiago, located near the top of the map, southwest of the city of Ilagan.
2. 6 – Philippine Sea, Sibuyan Sea, Visayan Sea, Bohol Sea, Camotes Sea and Samar Sea.
3. Mount Pinatubo, just north of Manila. The eruption produced around 10 billion tons of magma, and was one of only three Volcanic Explosivity Index 6 eruptions in the twentieth century.
4. Looc – 'cool' spelt backwards. The island of Tablas is located near the centre of the map, north of the island of Panay.
5. The Lubang Islands. Begin at Cape Encanto, located on the coast east of San Jose on the island of Luzon. Travel almost due west to Caiman (an alligator relative) Point, then travel south to the main airport at Olongapo. Found directly in a line due south of here are the Lubang Islands.
6. Roxas. Start at Cleopatra's Needle, just north of Puerto Princesa at the bottom-left corner of the map. The secondary road leads to the northern tip of the island, and from here the Linapacan Strait and the Mindoro Strait separate you from Mindoro Island, where the town of Roxas can be found. The other towns called Roxas are located on the north coast of the island of Panay, northeast of Cleopatra's Needle and north of Santiago on the island of Luzon.

60 JAPAN

1. Koga, located north of Tōkyō on the border to the northeast of Saitama.
2. 2,966 m, for Daibosatsu-rei, near the edge of the map west of Tōkyō.
3. I Nagara – when it gains an 'i' it becomes 'Niagara'. It is located across the bay from central Tōkyō, south of Ichihara.
   I Zushi, which becomes Sushi. It is located in the north of the peninsula in the Kanagawa district, south of Yokohama.
   I Warabi, located south of Saitama in the yellow, built-up area of Tōkyō, becomes 'wasabi'.
4. Midori. There are two locations with this name on the map, one near the top left and one southwest of central Tōkyō. Although it cannot be seen from the map, only the one in Gunma Prefecture, near the top left, is actually a city.
5. Tone, located on the Tone-gawa (river) northeast of Tōkyō.
6. Sakura – among other locations, there is a city of Sakura in Chiba Prefecture, to the east of Tōkyō.

61 CHINA

1. The Chang Jiang (Yangtze). This river is the longest in both China and Asia as a whole.
2. 'Hu' is the Chinese word for lake, as for example in 'Hong Hu', southwest of Wuhan. In the northern portion of the map are multiple bodies of water named 'shuiku', which is the Chinese word for reservoir.
3. Xiu Shui. Located in the southeast of the map, the two points given are both mountains.
4. Hanchuan. Named after the Han dynasty, Han Chinese is the dominant ethnic group in mainland China.
5. Henggang, south of the city of Huangshi and just across the administrative border. If you change the first 'g' to a space, it reads 'hen gang'.
6. Shaping. Start at Daye (which becomes 'day' without its final vowel) and head south. Head west from near Xintanpu, and continue to the broken motorway route just past Shaping (which can be a synonym for 'moulding').

62 INDIA AND SRI LANKA

1. 9 – Northern, North Western, Western, Southern, North Central, Central, Eastern, Sabaragamuwa and Uva.
2. Tamil, which occurs in the state name of Tamil Nadu, in the centre of the map. This language is part of the Dravidian family, and inscriptions in its ancient form have been discovered that are over 2,000 years old.
3. Salem, located near the centre of the map in Tamil Nadu.
4. Nazareth, located near the southernmost point of India.
5. Elephant Pass, a town on the north coast of Sri Lanka.
6. Madurai. Begin at Foul Point (a disallowed goal) on the east coast of Sri Lanka and travel around the coast to the northwest where you meet Adam's (first man) Bridge. The bold red line indicates a main road, which goes roughly northwest from the coast to Madurai. The black-on-yellow marker for Madurai indicates a population bracket of 1–5 million people.

63 UAE AND OMAN

1. Palm Islands, in Dubai. They are artificial islands in the shape of palm trees.
2. Sand. The legend shows that sand deserts are indicated by red dots and rocky deserts by small red open circles.
3. They are all oases.
4. The area is a dry salt lake or salt pan, according to the map legend.

5. Bāsa'īdū. The island is Qeshm.
6. Bat, Al-Khutm and Al-Ayn. Start at the Arabian Oryx Sanctuary, located in Oman near the bottom right-hand corner of the map. Go north to the Al Hajar' al Gharbī mountain range, where the highest labelled peak is 3018 m. The bold red line south of here indicates a main road, and travels northwest, passing the site of Bat, Al-Khutm and Al-Ayn on the left.

64 KYRGYZSTAN

1. The Union of Soviet Socialist Republics (also known as the USSR or Soviet Union).
2. Ysyk-Köl is a saltwater lake, indicated by the patterned infill, whereas Ozero Balkhash is freshwater in the section pictured. The latter is in fact semi-saline, and saltwater can be found in its eastern section.
3. Concord Peak (5469 m), located on the border of Afghanistan and Tajikistan.
4. Afghanistan
5. Pop, located just north of Qo'qon in the centre-west section of the map. Although the country name is not labelled on the map, Uzbekistan borders Kyrgyzstan to the west.
6. Qullai Karl Marks (6,726 m). Start at Lenin Peak on the border of Kyrgyzstan and Tajikistan, then head through the Tajik National Park. Qullai Karl Marks (Marx) mountain is located just to the south.

EUROPE

65 FINLAND AND RUSSIA

1. Tallinn, the capital of Estonia.
2. Pushkin. Alexander Pushkin (1799–1837) wrote many politically controversial poems, which led to his exile from Sankt-Peterburg (St Petersburg).
3. Polo. It can be found on a main road to the north of the map, near the eastern border of Finland with Russia.
4. Leningradskaya Oblast' and Respublika Kareliya. Ladozhskoye Ozero (Lake Ladoga) is the fourteenth largest lake in the world.
5. Novgorodskaya Oblast'. Start at Sankt-Peterburg (St Petersburg) and head east along the grey railway line to Volkhov. Follow the Volkhov river south to the lake Ozero Il'men' in the Novgorodskaya Oblast' district.
6. Narva Bay. Start at Malta in the east of Latvia (at the bottom left of the map), then travel east to the Russian border, and follow the border north to Narva Bay, situated in the Gulf of Finland.

66 NORWAY

1. Jotunheimen. This mountainous area is located towards the centre of the map, north of the 'R' in 'NORWAY.
2. Denmark. The islands are self-governing but rely on Denmark for some domestic responsibilities, such as defence and currency.
3. Gudbrandsdalen. The valley name can be seen in italic script following the course of the river that flows into Mjøsa lake.
4. I Roan, on the coast towards the northeast corner of the map.
   I Roald, the first name of author Roald Dahl, can be found just north of the coastal city of Ålesund.
   I Rindal, pointed at by the 'A' of NORWAY.
5. 3 – Sweden, Finland and Russia (at the far northeast)
6. Eina. The airport is near to Fagernes, and the river to its east is the Etna. This flows southeast into Randsfjorden. The main road to the east is indicated by a bold red line, where the town of Eina can be located.

67 UNITED KINGDOM

1. 3 – Exmoor National Park, Dartmoor National Park and Brecon Beacons National Park.
2. Glastonbury, located just northeast of the Somerset county name.
3. Yes Tor. This 619 m peak is just west of Chagford in Dartmoor National Park; Mumbles Head, at the end of Swansea Bay, just south

of Swansea itself; Simonsbath (Simon's bath), located near the source of the River Exe.
4. Flat Holm and Steep Holm, located in the Bristol Channel. 'Holm' is a mostly dialectal word for a small island.
5. The River Axe. Start at Rhoose airport, southwest of Cardiff, and fly to Bristol airport (located southwest of Bristol city). Head over the border between North Somerset and Somerset to the town of Cheddar. The river south of here is the Axe – axes can 'chop', and the river flows 'by' (water can also 'chop').
6. Beer, and all the locations start with the letter 'B'. Start at Bristol, where the Clifton Suspension Bridge is located. Travel to Bath, known as Aquae Sulis by the Romans. Head east-northeast to Box, then due south to Bovington Camp. Due west of here is Beer, just above the 'L' of Lyme Bay.

68 GERMANY

1. Speicherstadt & Kontorhaus District. This built-up area of warehouses and offices represents Hamburg's heritage as a port.
2. Kiel – pronounced in the same way as 'keel', part of the bottom of a boat. The German navy's Baltic fleet is traditionally based in Kiel, and the city has a long history of shipbuilding.
3. Weede, which can become 'weed'.
4. The Baltic Sea and the North Sea. Canals are marked by blue lines with short, perpendicular dashes.
5. The locations are:
   | Klütz, northeast of Lübeck
   | Plate, southeast of Schwerin
   | Bergen, near the bottom-left corner of the map
6. Lübecker Bucht (bay). The Elbe forms part of the border between Niedersachsen (not labelled) and Schleswig-Holstein states (two neighbours). Follow the border until you reach the border between Schleswig-Holstein and Mecklenburg-Vorpommern (not fully labelled), heading north from near Lauenburg. Your journey goes past the lakes of Schaalsee and Ratzeburger See, then past Lübeck (the founding city of the Hanseatic League, a commercial and defence alliance of cities in northern Europe) before ending at Lübecker Bucht.

69 THE ALPS

1. The Pennine Alps, on the border of Switzerland and Italy.
2. Mont Blanc, at 4,810 m – it is on the border of Italy and France
3. | Torino (Turin)
   | Reichenbach, near the top-right corner of the map (in the Conan Doyle story, Holmes and Moriarty fight next to the Reichenbach Falls)

Brig, just northeast of the VALAIS canton name.
4. Great St Bernard Pass (located east of Mont Blanc on the Swiss/ Italian border) and Col du Petit St-Bernard (located south of Mont Blanc on the French/Italian border), named after Saint Bernard of Montjoux, patron saint of the Alps and mountain activities.
5. Mercury. Start at Le Cheval Noir (a mountain which translates as 'the black horse'), located in the Savoie region of France. Follow the river directly north of the mountain past La Léchère, turn left onto the main road (bold red line) and head northwest to Albertville (Queen Victoria's husband was named Albert). West of here is a town named Mercury, which is also the closest planet to the Sun.
6. Val d'Hérens. Start at St Niklaus in the Valais district of Switzerland (St Nicholas being closely associated with Christmas and gift-giving) and follow the river north to Visp, then head west. Follow the motorway (a red and white line) which runs along the Rhône river until you reach the point between the towns of Montana (an American state) and Chalais (rhymes with Valais, the name of the province). The valley south of here is Val d'Hérens.

70 GREECE
1. Olympia, located near the coast on the west edge of the map. The site of the original Olympic Games is also marked on the map with the three-dot symbol indicating a site of specific interest.
2. Sparti (Sparta), located centrally in the bottom half of the map. Sparti led the Peloponnesian League to victory over Athens and the Delian League in the Peloponnesian War of 404 BC.
3. Korinthos (Corinth), located in the northeastern corner of the map. Corinthian columns are more ornate than other columns of the period, and can be recognized by their design of scrolls and acanthus leaves. The bible contains two letters from St Paul to the Corinthians.
4. Mycenae, located southwest of Korinthos (Corinth). The symbol for the site is just above the modern town of Mykinai.
5. Tripoli; Sapienza, an island in the southwest of Peloponnisos; Kalamata, on the south coast of the peninsula.
6. Argolikos Kolpos (also known as the Argolic Gulf). Start at Korinthos (Corinth) and head northwest along the motorway until you reach Velo (the French word for bicycle is 'vélo'), then head south passing Nemea to the west, the home of the Nemean lion which Hercules was challenged to kill by King Eurystheus. Continue south to Argos, which shares its name with the giant also known as Argus Panoptes. The bay south of Argos is Argolikos Kolpos, part of the Aegean Sea.

AFRICA

71 THE NILE

1. The Suez Canal (Suez is shown by its local form of As Suways on the map).
2. Cataract – specifically 1st Cataract (next to Aswān) and 2nd Cataract (next to Lake Nubia).
3. I Memphis, located to the southwest of Al Qāhirah (Cairo)
   I The Valley of the Kings, located just west of Al Uqşur (Luxor)
   I Abu Simbel Temple, north of Lake Nubia
4. Jabal Kātrīnā (Mount Catherine). St Catherine is said to have been sentenced to death using a breaking wheel covered in spikes (now known as a Catherine Wheel) and was carried to this mountain to be laid to rest.
5. Al Buḥayrah al Murrah al Kubrá (Great Bitter Lake), located on the Suez Canal, northeast of Al Qāhirah (Cairo).
6. The Red Sea. Start at the airport between the two borders of Israel at the northeast corner of the map. Travel through the Gulf of Aqaba past Abu Gallum Protected Area and Nabq Protected Area to Ra's Muhammed National Park. Continue southwest between the islands of Jazīrat Jūbāl and Jazīrat Shākir to the Egyptian coast, near the Red Sea Islands Protected Area.

72 BURKINA FASO AND GHANA

1. French. Both countries also have multiple local languages.
2. Niger
3. Parc National de la Fosse aux Lions, in the northern Savanes region of Togo.
4. Volta, all located in the southeast of Ghana
5. I Ashanti. The Ashanti Empire had an election system to choose its kings, and fought the British Empire during the 19th-century Anglo-Ashanti wars.
   I Prampram, found on the coast east of Accra.
   I Maritime, found in Togo.
6. A mole, in Mole National Park. Start at Bui National Park (just west of the GHANA country name) and follow the Mouhoun (Black Volta) river (the 'dark-sounding river') north along the purple international border until the thick red line of the main road crosses it. Follow the road eastwards, then head north from Bole to the town of Tuna, west of Mole National Park.

73 UGANDA AND TANZANIA

1. Mount Stanley, at a height of 5,109 m
2. | Rhino Camp, just south of Bidi Bidi
   | Puma, located in the Singida province, just south of the town of Singida
3. The Serengeti National Park, southeast of Lake Victoria in Tanzania
4. Lake Kyoga. Start at Kampala and travel east to Jinja where the Victoria Nile meets Lake Victoria. Follow the river north to Lake Kyoga.
5. Eldama Ravine. Start at Bunia and head south across the Ugandan border to Fort Portal ('a well-defended doorway'). If you drive south along the main road to Kasese and then follow the grey railway line, you pass by the UNESCO site labelled 'Tombs of Buganda Kings'. Keep going until you are near the eastern side of the map, where you find the airport southeast of Eldoret, and the settlement of Eldama Ravine ('chasm') nearby.
6. Lake Kitangiri. The world's largest unfilled caldera is the Ngorongoro Crater, located in Tanzania to the extreme right of the map. Head southwest to Lake Eyasi, then on to Lake Kitangiri.

74 SOUTH AFRICA

1. Xai-Xai, located on the coast of Mozambique near the eastern edge of the map
2. KwaZulu-Natal. The Zulu are the Bantu ethnic group
3. The border between Eswatini (Swaziland) and South Africa. Amsterdam (also the capital of the Netherlands) is located to the west of Eswatini (Swaziland).
4. | Sun City, located on the border of Pilanesberg National Park, northwest of Pretoria (Tshwane).
   | Lady Grey tea is flavoured with citrus and bergamot, and Jane Grey was deposed when the Privy Council changed their allegiance to Mary Tudor (Mary I)
   | Golden Gate Highlands National Park, which shares part of its name with the Golden Gate Bridge.
5. Winter, in the town of Winterton. Start at Maletsunyane Falls, a 192-metre-high waterfall. From here, head northeast to Champagne Castle, a mountain on the border of Lesotho and South Africa. Winterton is just north of here on the main road.
6. Pinetown/Cedarville. Start at St Lucia Estuary, located some distance due south of Maputo, and head south to Durban. Pinetown ('coniferous') is located directly next to and northwest of Durban. 'Pinetown' is a tree and a word for a settlement stuck together, which is a property shared by Cedarville (Cedar + ville).

NORTH AMERICA

75 ALASKA, USA

1. Bristol, in Bristol Bay
2. The Kodiak bear. The 'K' of Kodiak Island just appears at the edge of the map
3. Hamilton, at the top left of the map. Alexander Hamilton was the first Secretary of the Treasury; Farewell, just north of Mount Hesperus in the northeast section of the map; Kanakanak. Nushagak Bay is located at the north edge of Bristol Bay, just inland from Cape Constantine.
4. Mosquito Mountain. Start at the Yukon river delta towards the northwest of the map, then travel along the river past Pilot Station (a 'town for aviators') to Holy Cross (a 'religious burden'). Mosquito Mountain is east of here
5. Kisogle Mountain. Start at Walrus Islands towards the north of Bristol Bay and head north to Twin Hills, then travel around the coast in a westerly direction to Goodnews Bay. If you follow the Goodnews river inland, Kisogle Mountain is on the left.
6. Battle Lake. Start at King Salmon river near the north of the Alaska Peninsula, and follow it to its source near Mageik Volcano. Head northeast across this and Mt Katmai to Snowy ('white') Mountain. Finally, head north to Battle Lake ('a place of suggested conflict').

76 CANADA

1. | Hannah, just on the US side of the US–Canada border near the bottom left of the map.
   | Reykjavik (also the capital of Iceland) can be found near Lonely Lake ('solitary lagoon') on Lake Manitoba.
   | Badger, located just west of Roseau.
2. Rivers. Carman is located southwest of Winnipeg, while St Ambroise is north of this point on the edge of Lake Manitoba. There are three railway lines (marked by grey lines), and four rivers between these two points. There is also a railway and river at Carman itself.
3. Drunken Point. Following the grey line from Stonewall, situated just north of Winnipeg, the end of the line is Arborg, and due east of here is Drunken Point on Lake Winnipeg.
4. Moose Island, Reindeer Island, Deer Island and Elk Island are all named after animals which belong to the Cervidae family.
5. Berens Island. Catch a flight from the airport in Winnipeg and head north over Wicked Point and McBeth Point (similar to the title of the play 'Macbeth', with its famous line, 'Something wicked this way comes') and land at Berens River airport. From here, looking west, the island opposite you is Berens Island, situated in Lake Winnipeg near the mouth of the Berens river.

6. Dancing Point. Starting at Winnipeg, take the train northwest towards Moosehorn ('antler'). From here, carry on along the road in the same direction until you reach the river that leads into Lake St Martin (named after a French saint), and travel across them both to the Dauphin river (the dauphin was historically the eldest son of the King of France) at the northeastern shore of the lake. The river empties into Sturgeon Bay (a 'fishy harbour'), and following the coastline northwest will lead you to Dancing Point (where you can 'show your moves').

77 NEW YORK, USA
1. The Statue of Liberty.
2. Atlantic County and Ocean County, making Atlantic Ocean.
3. | Pearl River, just on the border of Rockland County and Bergen County. Pearl River is the name of the town – although there is no river of that name nearby!
   | Deal. On this map, Deal is on the coast due south of New York City – there is also a Deal in Kent, England.
   | Bergen County, due north of New York City.
4. Oyster Creek. Start at the town of Mystic Island (the 'supernatural isle') just east of Atlantic County and take the secondary road north. From here turn left onto the main road before Tuckerton, left onto the motorway before New Gretna, left to leave the motorway and left again at Smithville. The road leads to Oyster Creek, which is within the wildlife reserve – and across from where you began.
5. Pleasantville. Starting in Vineland (where you might grow grapes) towards the bottom of the map, follow the train line, marked by a grey line, north to Malaga, which shares its name with the Andalusian city in Spain. The main road travelling southeast crosses through Buena (the Spanish word for 'good') and on to Pleasantville ('agreeable town'), and the first one you reach as you enter the yellow section that indicates a built-up area.
6. The Raritan. Stockholm (the capital of Sweden) is in Sussex County in the north, Roosevelt (the wartime President) is in the west of Monmouth County. The airport next to Valhalla (the Viking resting place) is at the northeast of the map north of Greenwich, and Princeton (in Mercer County) is the university with Tigers (its athletic nickname). The two lines intersect just south of the Raritan river, which flows into the Atlantic.

78 FLORIDA, USA
1. Satsuma, at the centre-top of the map, south of Palatka.
2. Frostproof, near the centre of the map, above the 'O' in FLORIDA.
3. | Melbourne (east coast, central).
   | St Petersburg (west coast, central)

| Venice (west coast, south of St Petersburg)
| Naples (south of Bonita Springs).
| Venus (west of lake Okeechobee).

4. At the centre of the peninsula at the top of the map.

5. Pasco. The triangle connects Hudson, Inverness and Zephyrhills, all located in the area north of Tampa and west of Orlando. The only town with five letters in this triangle is Pasco.

6. Holopaw. Start at Crows Bluff, head south over Orlando (as in Orlando Bloom) to St Cloud (a 'canonized mist'). From here, the main road travelling southeast towards the coast reaches Holopaw first.

79 GUATEMALA AND MEXICO

1. The relative closeness of the contour lines and their 1000 m intervals show that the sea floor drops off sharply to form a deep trench just off the coast of Guatemala and the Mexican state of Chiapas. It is known as the Middle America Trench.

2. | Tabasco, found on the western side of the map.
   | Progreso. The words 'Ordem e Progresso' appear on the Brazilian flag, which translate into English as 'order and progress'.
   | Ambergris Caye. Ambergris initially has a sewer-like smell, but loses this as it ages and is used to stabilize other scents.

3. It becomes the border between Mexico and Guatemala. Starting at Mérida (just south of Progreso) the train line, marked with a grey line, travels to Tenosique on the border of the states of Chiapas and Tabasco. This is on the river Usumacinta, which when followed upstream (i.e. southeast) becomes the course of the border between the two countries.

4. Tulum. Start at Isla de Cozumel, just off the east coast of the Mexican state of Quintana Roo. The town in the south is Cedral, and due west of here is the historical site of Tulum, marked by three dots.

5. Southwest. Starting at the Parque Nacional El Tigre (the striped animal, since 'tigre' translates to 'tiger' in English) in northern Guatemala, travel cross-country to the town of La Libertad – the Spanish word for liberty. From here take the road northeast to Belize City. Southwest of here is Belmopan, the capital of Belize.

6. 1,060 m. Start at the Volcán de Tajumulco in Guatemala at a height of 4,210 m, which is the highest volcano in Central America. The river just north of here flows into Mexico and becomes part of the Presa de la Angostura, which shares its name with Angostura Bitters. At the northern end of this reservoir (the English meaning of 'presa') the river Mezcalapa flows to the city of Tuxtla Gutiérrez. Due north of here is the volcano El Chichónal, with a height of 1,060 m.

80 COLOMBIA

1. The Pacific Ocean, seen on the central western side of the map.
2. The Parque Nacional La Macarena, Cordillera Macarena and La
   Macarena can all be found west of the 'GUAVIARE' department
   name. The 'Macarena' is a Spanish song with an accompanying
   dance.
3. | Santander
   | El Dorado, located in the southeastern corner of the map. This
   poem is about a knight who fruitlessly searches for the mythical city
   of El Dorado.
   | Parque Arqueológico San Agustín, a UNESCO World Heritage
   Site, shown by the three dots just west of Florencia in southern
   Colombia.
4. Magdalena. A line drawn between Bogotá and Armenia (a
   'Caucasian country') crosses the River Magdalena, which continues
   north towards the department of Magdalena, to which it lends its
   name.
5. 5,399 m. Start at the train station in Cesar (sounding like the title
   'Caesar', given to Roman emperors) and take the train route (marked
   by a grey line) south. The right-hand fork will take you to Medellín,
   the second-largest city in Colombia by population. From here,
   looking southeast, the nearest volcano (marked by a red triangle) is
   the Nev. ('Nevado' - the Spanish word for snowy) Del Ruiz with a
   height of 5,399 m.
6. At Puerto Rico, in the foothills of the Cordillera Oriental. The
   settlement of Argentina (the second-largest South American country)
   can be found on the Caguán River, near the bottom of the map.
   Following the river north and using the left fork after the L. de Chaira,
   you come to Puerto Rico (sharing its name with the island of Puerto
   Rico, an unincorporated US territory) which is in the foothills of the
   Cordillera Oriental. The name of the mountain range stretches across
   much of the southwest portion of the map.

81 BRAZIL

1. Ilha Grande.
2. | Olímpia. Southwest of the city of Barretos in the São Paulo province, this town sounds similar to the site of Olympia in Greece, the location of the first Olympic Games.
   | Orizona. It is one letter away from Arizona, USA and is located southwest of the mountain with a height of 1,233 m, south of Brasília.
   | Colombo, located next to Curitiba in the southwest corner of the map.
3. Goiânia. Frutal (one letter away from both 'frugal' and 'brutal') can be found on the main road just north of the city of Barretos, close to the Grande river. Following the main road north will lead you to Goiânia, whose town stamp indicates a population of between 1 and 5 million people.
4. São Francisco. The main road east of Uberlândia leads to Pirapora, which starts with 'Pi' (an 'irrational number'). Following the river northwards from here you eventually reach São Francisco, which is similar in name to the Californian city of San Francisco, where the Golden Gate Bridge is located.
5. Quebra Anzol. Start from Franca, just south of the Grande river near the centre of the map. Cross the Grande (Portuguese for 'great') to Sacramento (the state capital of California), then go north to the lake next to Nova Ponte, out of which the Quebra Anzol flows, in the east corner.
6. A jaguar, in the town of Jaguariaíva. Start at Curitiba, indicated to have more than 1 million people by the black square inside a yellow square. Directly north of here, there is a grey railway line which intersects a purple province boundary in the Serra Paranapiacaba mountain range. Follow the path of both the river and the purple boundary to Itararé, then head west to Jaguariaíva.

82 TIERRA DEL FUEGO

1. Lago Buenos Aires (L. Buenos Aires on the map) can be found just east of the border between Chile and Argentina, and just south of the administrative division near the top of the map (marked by a purple line). The lake is known as Lago General Carrera on the Chilean side (L. Gen. Carrera on this map).
2. | One – just south of Rio Gallegos. Main roads are marked on the map by a bold red line. There are other main roads that reach the border, but which become secondary roads – according to the map legend – as they cross into the other country.

Caleta Olivia. El Pluma is in the north of Santa Cruz province. The main road east will take you through Colonia Las Heras. At the junction by Pico Truncado, the only main road option is the one that leads northeast to Caleta Olivia.

3. Parque Nacional Bernardo O'Higgins, in Chile. Volcán Lautaro, at a height of 3,623 m, is marked by a red triangle.

4. Bella Vista. The Cueva de las Manos (translated as 'Cave of Hands') is marked by three dots on the map, which signify a site of specific interest, and is found in northwestern Santa Cruz. Due south of here – according to the curved lines of longitude – is Bella Vista, whose name translates as 'beautiful view'.

5. The river Deseado. The only marked volcano in Argentina on this map is the Vol. Elisabeth, just below the Santa Cruz province name. Sarmiento is the town to the north of here which is an anagram of 'iron mates', and between these two points is the river Deseado.

6. San Gregorio. The world's southernmost commercial airport is in Ushuaia, at the bottom of this map. Following the border north you reach the coast of Tierra del Fuego (a 'fiery' island, as fuego translates to 'fire' from Spanish) just below Punta Catalina – and due west of here is San Gregorio, which shares much of its name with the Gregorian calendar – the most widely-used calendar in the world.

OCEANS AND POLES

83 MARIANA TRENCH
1. Bikini – on the Bikini Atoll.
2. The Tropic of Cancer.
3. | Yellow Sea, located between South Korea and China.
   | Mapmaker Seamounts, on the eastern side of the map. Seamounts are underwater mountains, usually formed by volcanic activity.
   | New Caledonia, in the southeast of the map. Caledonia is the archaic name for the country of Scotland, given to it by the Roman Empire. 'New Caledonia' is therefore both old and new.
4. The Torres Strait, between Australia and the island of New Guinea. The Torres Strait Islanders have a distinct culture with some Australian and some Papuan elements.
5. Challenger Deep, the deepest recorded point in the Earth's ocean floor, at 10,920 m below sea level. To the south coast of Hokkaido follow the paths of the Japan Trench, the Izu-Ogasawara Trench and the Mariana Trench southwards until you reach the ocean depth marker. This is Challenger Deep (a 'profundity' that has 'thrown down the gauntlet').
6. Ferdinand Magellan – at the Magellan Seamounts. The Solomon Islands are the archipelago named after the wise king, and due north of this along the line of longitude is Wake Island – the 'island to rise and shine'. West and slightly south of here are the Magellan Seamounts, which share their name with Ferdinand Magellan, the Portuguese explorer.

84 NORTH POLE
1. Alert. It can be found at the very top of Ellesmere Island.
2. Longyearbyen, Svalbard. It is represented by a yellow square.
3. Qaanaaq
4. | Lincoln Sea. Abraham Lincoln was the sixteenth US President.
   | Kaffeklubben Ø, to the north of Greenland. It is reputed to be named after a regular informal gathering that took place at the University of Copenhagen, attended by Lauge Koch, who named the island.
   | Île de France. Located on Greenland's eastern coast, this uninhabited island shares its name with the administrative région of France that contains the capital city, Paris.
5. 4 – Canada (although not labelled on the map, Ellesmere Island is

part of Canada), Denmark (Greenland), Norway (Svalbard) and the USA (at Thule Air Base).

6. Qimusseriarsuaq. It can be found just below the Thule Air Base, off the west coast of Greenland.

85 ANTARCTICA

1. Mount Vinson, at 4,892 m high. It can be found in the mountain range labelled 'Ellsworth Mountains', near the centre of the map.

2. 2017. The measurement is shown by a dotted line, and a description just below the label for 'Weddell Sea'.

3. | Ketchum Glacier. It is one letter away from 'ketchup' and can be found on Palmer Land near the 75° line of latitude.
   | Cape Mackintosh. It is located on the Antarctic Peninsula, on the east coast of Palmer Land.
   | Rupert Coast. It is located at the bottom of the map. The fictional bear is Rupert Bear.

4. | Halley VI, located near the top of the map on the Brunt Ice Shelf. 'Halley's comet' and Halley VI are both named after Edmond Halley, the English astronomer who lived from 1656–1742.
   | Belgrano II. The *ARA General Belgrano* is the name of the ship. It is named after Manuel Belgrano, a key figure in the campaign for Argentinian independence from Spain.
   | Queen Elizabeth. Queen Elizabeth Land is near the centre of the map, between the Academy Glacier (a chilly place of study) and Skytrain Ice Rise (an elevated locomotive).

5. North America – although the line crosses some countries in Central America, they are continentally part of North America.

6. Thurston Island. Start at Bear Peninsula (part of the Ursidae family) and travel along the coast, around Thwaites Glacier Tongue, to the Abbot Ice Shelf. Cross over the shelf to reach Cape Flying Fish (part of the Exocoetidae family), located on Thurston Island.

THE TIMES WORLD ATLAS CROSSWORD

ACROSS

1 Arab
5 Vienna
6 Superior
10 Egypt
14 Ohio
15 Reno
17 Jakarta
21 Dane
22 Bucharest
25 Aleppo
26 Siam
27 Atacama
29 Niger
30 Idaho
33 Hun
35 United Kingdom

36 USA
38 Virginians
40 Cannes
41 Alabama
43 Belfast
45 Miami
46 Iran
47 EEC
48 Aden
49 Etna
50 Colca Canyon
51 Pretoria
53 Tuvalu
55 Mali
56 Nile
58 Zagreb

60 Ural
61 Ria
63 Uruguay
64 Namib
67 Cork
68 Rio
69 Lagos
70 Asti
73 Niue
75 Iowa
76 Australia
77 Oman
78 Pitcairn
79 Alp

DOWN

2 Benin
3 Georgia
4 Faro
6 Suez
7 Patagonia
8 Rabat
9 Oklahoma
11 Gulf
12 Tepui
13 Guam
16 Lebanon
18 Kandahar
19 Riga
20 Agra

22 British Columbia
23 Cardiff
24 Addis Ababa
26 Sahara
28 Abuja
31 Davos
32 Tonga
33 Han
34 Accra
37 Qatar
39 Asia
42 Banjul
44 Libya
45 Montana

50 Cairngorms
51 Peru
52 Espana
54 Amur
57 Eyre
59 Africa
62 Andes
65 Milan
66 Rome
70 Avon
71 Aztec
72 Walers
74 Iraq
77 Oslo

ACKNOWLEDGEMENTS

Many thanks to Laura Jayne Ayres and Elizabeth Crowdy for all their help in creating this book.

Also to Karen Midgley for her editorial skills, Gordon MacGilp for his design work, and Jethro Lennox and Lauren Murray at HarperCollins Publishers for guiding the book through the whole process.

For their exhaustive cartographic knowledge, thanks to Jim Irvine, David Mumford, Kenneth Gibson and Craig Asquith at HarperCollins Publishers.

Gareth also wishes to thank his wife, Sara, for her continuing support.